92914

Millions of Monarchs, ..ches of Beetles

GILBERT WALDBAUER

Millions of Monarchs

HOW BUGS FIND STRENGTH IN NUMBERS

Bunches of Beetles

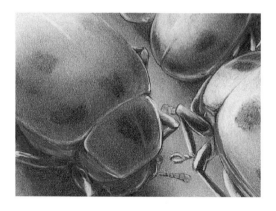

HARVARD UNIVERSITY PRESS

CAMBRIDGE, MASSACHUSETTS, AND LONDON, ENGLAND

First Harvard University Press paperback edition, 2001

Illustrations by Katherine Brown-Wing

Library of Congress Cataloging-in-Publication-Data

Waldbauer, Gilbert.
Millions of monarchs, bunches of beetles : how bugs find strength
in numbers / Gilbert Waldbauer.
p.   cm.
Includes bibliographical references.
ISBN 0-674-00090-0 (cloth)
ISBN 0-674-00686-0 (pbk.)
1. Insects—Behavior.   2. Social behavior in animals.   I. Title.
QL496.W36   2000
595.7156–dc21   99-42453

To my beloved daughters,
Gwen Waldbauer Rose and Susan Waldbauer Yates,
whose love and encouragement support me beyond measure

Contents

Strength in Numbers

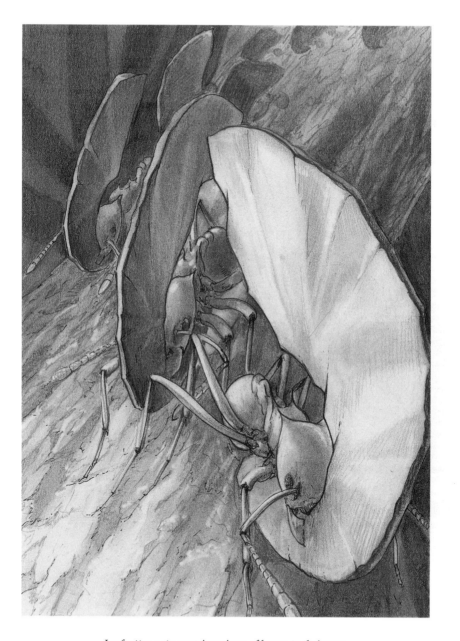

Leafcutter ants carrying pieces of leaves to their nest

⚙ ⚙ ⚙   Two especially memorable experiences with insects stick in my mind. Early one morning, when I was still a teenager in Fairfield County, Connecticut, I watched a procession of eastern tent caterpillars leave their tent in the main fork of a wild black cherry sapling, and crawl along branches and twigs to a spray of leaves on which they commenced to feed. The caterpillars moved nose to tail in single file. The procession was orderly, one might even say stately. It reminded me of the parade that marched down Main Street in nearby Bridgeport every Memorial Day. Years later, long after I had become a professional entomologist and while I was on sabbatical leave in Colombia, I watched, for the first time, a column of leafcutter ants move over the ground to the huge mound of soil that marked the entrance to their underground nest. The column was long. "Empty-handed" workers moved out, and returning workers delivered inch-wide pieces of green leaves that they held in their jaws and carried over their heads like parasols. I knew from my reading that the nest was inhabited by a highly organized society of agriculturists that use leaf fragments to make a rich mulch on which they cultivate the fungus that is their only food. These two processions impressed me because the caterpillars and the ants—which most of us think of as *lowly* insects—quite obviously operated as organized groups that acted in concert.

People pay particular attention to insects or other animals that occur in groups. Who would not be awed by a swarm of desert locusts so immense that it blocks the sun? Early visitors to the western plains of the United States were so impressed by the huge herds of bison they saw that they sent home excited reports of these vast assemblies. Who would not marvel at a group of beetle larvae (the immature, growing stage) that join together to repel predators or at a group of caterpillars that cooperate to build a silken tent to live in? Scores of writers have extolled the virtues of the social honey bees. And how many children—adults, too—have been fascinated by the many organized activities that go on in a glass-sided ant farm?

Entomologists interested in insect groups have focused most of their attention on the highly organized societies of ants, bees, wasps, and termites, probably because of the obvious and inevitable comparisons between them and human society. Such comparisons have been made for millennia. In the Bible we read (Proverbs 6:6), "Go to the ant thou sluggard; consider her ways, and be wise." In *King Henry V*, Shakespeare wrote, "for so work the honey bees, creatures that by a rule of nature teach the act of order to a peopled kingdom."

Honey bees and ants, such as my leafcutters, as well as termites and bumble bees and some wasps, form the most complex and cohesive of all insect societies. Their study is an immense field, only touched upon in this book, but thoroughly covered by several authoritative and readable books: among them two on social insects in general by Edward O. Wilson, two on ants jointly written by Bert Hölldobler and Wilson and another by A. F. G. Bourke, two on honey bees by Karl von Frisch, a more recent one on honey bees by Thomas Seeley, one on all social bees by Charles Michener, and another on social wasps by Kenneth Ross and Robert Matthews.

Less attention has been paid to more or less unorganized groups and simple societies such as that of the eastern tent caterpillar, probably because the comparison with human society is less obvious. But from a biological point of view, these groups, the main focus of this book, are as interesting as the more complex societies, and there is much we can learn from them. Group living at any level is important in the ecological scheme of things, because it enhances survival.

⁂ ⁂ ⁂ In the 1920s in a tropical forest in British Guiana (now the independent nation of Guyana), Maud Haviland watched a group of tiny leaf beetle larvae ward off predators by "circling the wagons." The beetle larvae feed in a compact cluster on the upper surface of a leaf of a cecropia tree, and, according to Haviland, they form a tight circle with their heads all directed inward and their armored tail ends thrashing to and fro at the periphery of the circle as they nibble on the leaf. The last segment of a larva's body is armored, broadly expanded to form a hard, impenetrable, shovel-shaped shield. The chief enemies of the leaf beetle larvae, insect-eating stink bugs, lurk nearby, waiting for an opportunity to lunge forward and spear a larva on their long, thin, piercing-sucking beaks. As Haviland put it, "As long as the circle of [shields] is unbroken, the bug stands little chance, for his [beak] cannot penetrate their polished armour and he cannot reach the soft bodies beyond." But as the larvae consume more and more of the leaf, they must back away from the center of the circle, thus widening it and leaving broader gaps between their armored shields, although the circle remains intact to the end. Eventually a stink bug or two penetrates the defensive formation and spears a larva, but most of the larvae survive because they are almost fully grown and ready to escape the stink bugs by descending to the soil to metamorphose to the pupal stage of the life cycle, the transformation stage that bridges the larval and adult stages. Haviland did not mention the name of these beetles, but according to Pierre Jolivet and Trevor Hawkeswood, leaf beetles of the genus *Coelomera* occur in Guyana and the larvae are known to use this defense.

High in the Canadian Arctic, about 5,000 miles north of Guyana, a herd of muskoxen grazes on a windswept slope on Ellesmere Island, eating grasses and other tundra plants such as the prostrate Arctic willow, which the oxen expose by scraping away the thin covering of snow with their hooves. When a pack of wolves appears, the great, shaggy grazers—some weigh over 700 pounds—quickly come together in a defensive formation. According to an idealized account that has been passed down from author to author, the muskoxen use a ring defense much as do Maud Haviland's beetle larvae, gathering in a tight circle with the adults facing outward, and the calves inside the circle or huddled between the adults. But Anne Gunn, a biologist who studies muskoxen in the Canadian Arctic, told me that the defense is seldom that orderly. In response to a pack of wolves, muskoxen do gather in a tight bunch with the calves in a protected position, but the adults tend to mill about and jostle each other and seldom form a circle. Their massive heads are lowered and their sharp, down-curved horns—those of the cows are almost as formidable as those of the bulls—are formidable weapons, ready to gore or rip open any wolf that comes too close. From time to time, a bull dashes out to attack one of the wolves, trying to kill or disable it with a sweep of his horn. Although a pack of wolves sometimes manages to kill a lone muskox, the defensive formation of a herd is virtually impregnable to attack, and the wolf pack eventually becomes discouraged and moves away.

Maud Haviland's leaf beetles and muskoxen have discovered—quite independently of each other—that there is strength in numbers, that an individual can benefit by belonging to a group and participating in group actions. The similarity in the defensive tactics of these two unrelated animals, one a tiny insect and the other a huge mammal, is a striking example of convergent evolution, of different creatures independently evolving similar solutions to similar problems, in this case the ring defense or some approximation of it.

Other kinds of animals also form defensive groups when threatened. Among them are the nasute soldiers of certain species of termites, which form rows of protective flanks on either side of the columns of workers that leave the nest to forage for food. The head of a nasute, a name derived from the Latin for large nose, is drawn out into a long, forward-pointing tube that can squirt a toxic and very sticky substance onto any insect or other creature that threatens the column of foragers.

The predaceous larvae of owlflies (not really flies but relatives of the antlions and aphidlions) stay together in a defensive huddle during the first few days of their lives, forming a tightly packed cluster on a twig near the egg mass from which they hatched. Charles Henry described how they hang head downward on and above their empty egg shells, "overlapping like shingles in such a manner that only the heads and jaws of individual larvae are visible." They capture and eat small insects such as fruit flies and midges that happen to come close. But in response to ants or other predators, they raise their heads and rapidly snap their long, sickle-shaped jaws. After about a week, the owlfly larvae descend to the ground, separate, and henceforth lead a solitary existence as they sit hidden in debris waiting to ambush passing insects.

Young Colorado potato beetle larvae cluster together more or less tightly as they feed on a leaf, and as E. R. López and his coauthors noted, they defend themselves by moving closer together when approached by flies that can infest them with lethal parasitic maggots. At the approach of one of these flies, which the beetle larvae can detect from a distance of about 20 inches, they rear up on their hind legs in unison and flail their front legs, thereby fending off the flies about 45 percent of the time.

If the leaf beetle larvae in Guyana are ever outmaneuvered by some predaceous insect that evolves a way of broaching their defensive formation, they may be doomed to ultimate extinction. They may, however, evolve some new way to defend themselves. But if the hunted, the leaf beetle larvae in this case, evolve new ways to defend themselves, the hunters, in this case stink bugs or other predaceous insects, must evolve new ways to circumvent the new defenses of their prey, switch to other prey, or perhaps even face starvation and extinction. Such escalating evolutionary arms races between hunters and hunted, still going on today, are pervasive chapters in the history of the evolution of life on earth.

⚙ ⚙ ⚙ Evolution is the central and unifying concept of biology, the science of life, the science through which we seek to understand ourselves and our fellow creatures, to know where we came from, what we are, and how we are inextricably bound to all other life on earth. We cannot completely understand ourselves unless we understand nature, and we cannot truly comprehend nature unless we understand evolution. How, then, does evolution work?

The driving force of evolution is Charles Darwin's concept of natural selection, also known as the survival of the fittest. The success, or fitness, of an animal, not necessarily achieved by proficiency with fang and claw, is measured by the number of progeny that it leaves behind. Natural selection produces new species much as livestock breeders produce new breeds by artificial selection, by choosing only animals with desirable, heritable traits, traits that are fixed in the genes, to be the parents of the next generation. Similarly, natural selection tends to eliminate poorly adapted individuals and to favor the survival of those that are best adapted to their environment because they are better able to avoid its hazards or take advantage of the opportunities it offers. For example, a butterfly with a longer than usual tongue can sip from the blossoms of more kinds of nectar plants than can a shorter-tongued individual of the same species and may, therefore, be better fed than a short-tongued individual and thus likely to produce more progeny; a more deceptively camouflaged individual is less likely to be noticed by a predator and, therefore, more likely to survive to become a parent than is a less well camouflaged individual of the same species. In this way, nature selects the parents of the next generation, and if the advantageous characteristics of the parents are heritable, they will be passed to their children, grandchildren, and on into the distant future. Even if the selective advantage is small, a favorable characteristic will, given enough time, replace a less favorable one. There is no shortage of new heritable traits. They constantly arise as mutations caused by certain chemicals, radioactivity, ultraviolet light, cosmic rays, or by intrinsic factors in the genetic material itself. Some mutations are favorable and some are not. Natural selection tends to eliminate the unfavorable ones and generally preserves the favorable ones.

The fossil record, which tells the story of evolution, is incomplete. In other words, there are gaps that are often referred to as "missing links." If you view a fossil record as a chain, you can see that there are many missing links but that there are also many links that are not missing, often enough of them to form a respectable segment of chain, a complete or nearly complete and convincing record of how an organism evolved to become a new species.

Scientists refer to the *theory* of evolution, but they don't use this word in its everyday sense. To a scientist, a theory is a concept that is strongly supported by evidence and about whose validity there is very little doubt, such as the theory of gravity. A hypothesis, on the other hand, is an idea that is not as well supported by data, but that *can be tested* by experimentation or observation and that may some day be so strongly supported by data that it advances to the status of a theory. The proviso that it must be possible to test a hypothesis by experimentation and observation is all important. No scientist can take seriously a hypothesis or other form of explanation that is not testable. One of the great strengths of science is that scientists are—or should be—both conservative, cautious about accepting new theories, and open-minded, willing to consider new data. They recognize that their understanding of almost any matter can be improved through more experimentation and observation, as is the case with our comprehension of evolution. There is no doubt that living things evolved and continue to evolve, but much remains to be learned about the details of the process.

Three levels of natural selection are recognized: *selection on the individual,* as just described; *kin selection,* which enhances an individual's *inclusive fitness;* and the controversial but reasonable concept of *group selection,* selection that operates on the group as a whole rather than only on the individuals in a group.

Kin selection, a brilliant concept first expounded by William F. Hamilton, recognizes that any individual that sacrifices itself for the good of its relatives will enhance its own inclusive fitness by contributing to the survival of those relatives, which contain many of the individual's genes. A well-known example is the honey bee worker that dies defending the colony against a marauding mammal, injuring herself mortally when her barbed stinger penetrates the marauder's skin and tears away from her own body. As you know, the fitness of an individual is usually considered to be a matter of how many offspring it leaves behind. Offspring are the bearers of their parents' genes, the very genes that programmed the anatomical, physiological, and behavioral characteristics that promoted the survival and reproductive success of their parents, and it is through these offspring that these genes will be passed on to future generations. But, as Hamilton pointed out, the ultimate measure of an individual's evolutionary success is its inclusive fitness, which is measured by the survival of its genes, whether they are contained in its own body or in the bodies of relatives. Each parent gives each of its offspring half of its genes: thus brother and sisters share half of their genes; first cousins share one eighth of their genes with each other; and one quarter of a grandparent's genes appear in each grandchild.

As Elliott Sober and David Sloan Wilson pointed out in *Unto Others,* the concept of group selection has its roots in the writings of Charles Darwin, who invoked it sparingly and with a critical eye. But Darwin's successors "were less abstemious, invoking the process [of group selection] widely and often uncritically." V. C. Wynne-Edwards, for example, argued that animals come together in groups so that they can judge the size of their population, and if it is too large they can, as Sober and Wilson expressed Wynne-Edwards' idea, "restrain themselves from consuming food and reproducing, so that the population can avoid crashing to extinction." A backlash against this and other obviously improbable ideas gave the concept of group selection an undeserved bad name. Nevertheless, some observations and experiments with organisms ranging from bacteria to vertebrates cannot be explained without invoking group selection. The antipathy to group selection is gradually disappearing, but still persists in some quarters.

A good example of group selection occurs in the fungus-growing desert leafcutter ant of Arizona, *Acromyrmex versicolor.* A new colony of this species is founded by a group of several *unrelated* queens, usually in the shade of a tree that is likely to attract the foundresses of several other colonies. One of a colony's founding queens acts as the forager that collects the plant material on which the ants grow the fungus that is their only food. This queen becomes an ever more efficient forager with experience, and can be said to act "altruistically" for the benefit of the colony as a whole, because she leaves the nest and is thus more exposed to dangers such as predators than are the queens that stay in the nest to tend the fungus garden. But her specialized prowess determines the rate at which the new colony develops—a crucial factor because only one of the colonies under the tree survives, the one that first produces a contingent of adult workers. These first workers destroy the other colonies under the tree, raiding them to steal their brood, larval and pupal workers, which they raise as their own to augment their force of workers. Kin selection cannot explain the cooperation between the founding queens of a colony because, as genetic tests have shown, they are not related to each other. Group selection seems to be the only plausible answer.

⊕ ⊕ ⊕   Insect groups may be relatively small, consisting of anywhere from less than a dozen individuals, as in a feeding group of jack pine sawflies, to a hundred or more, as in a family of aphids or a sleeping aggregation of wasps. Others may be huge, including thousands or even millions or billions of individuals, as do swarms of periodical cicadas or the grasshoppers known as migratory locusts.

A group may form only because its members grow and develop synchronously, as do some mayflies, and thus appear together at the same time in the same place, or it may form because, as is the case with chinch bugs, a number of individuals gravitate to the same resource. Groups also form because members of the same species attract each other, as when thousands of fireflies gather in the same tree to attract others by flashing in synchrony. There may be little or no interaction between the members of a group, as is the case with mayflies and chinch bugs; there may be moderately complex interaction, as in a subsocial group of tent caterpillars; or there may be highly complex and all-pervasive interaction, as in the eusocial ("truly social") colonies of termites, ants, honey bees, bumble bees, and some wasps.

Even groups that are seemingly incidental and have little or no internal organization may, nevertheless, be favored by natural selection because they increase the probability of finding a mate or reduce the likelihood that an individual will fall to a predator because it is less vulnerable in a crowd. Thus if the individual benefits from being a member of an unstructured aggregation, natural selection may tend to perpetuate and enhance the formation of such aggregations, not by favoring individuals that are attracted to each other but, rather, by tightening the response to an outside synchronizing stimulus by eliminating stragglers that respond either too early or too late.

Groups of aphids consisting of from a few dozen to a hundred or more siblings form on their food plants because newly born aphids do not move far from the mothers that gave birth to them. In some species, an aphid that is attacked by a predator gives its nearby siblings warning by emitting a special odorous pheromone—a chemical signal released by one member of an animal species that provokes some physiological or behavioral response in another member of the same species.

Immature mayflies lead independent lives in the muck at the bottom of lakes or rivers, but because of their tightly synchronized emergence from the water, they occur together as swarms of flying adults, often in astronomical numbers. Their emergence is so synchronized because every individual is triggered to molt to the adult stage and leave the water on the same day and at the same time by the same extrinsic environmental stimulus or stimuli, perhaps by the length of the day or the temperature of the water. Dead mayflies sometimes accumulate under lights along the shores of the Great Lakes or the banks of midwestern rivers in great piles that may be several feet deep. The unfortunate insects are irresistibly drawn to the lights during the night and beat themselves to death.

Monarch butterflies are solitary during much of their adult stage and always as eggs, caterpillars, and pupae, but they survive the winter by migrating to form huge aggregations of adults clustered in trees in the mountains of Mexico or along the coast of California. In the 1970s a 5.5 acre overwintering site in Mexico was found to contain about 22.5 million monarchs, clustered so tightly that they obscured the foliage and made the trees on which they rested look orange rather than green.

The adult 17-year periodical cicadas—sometimes incorrectly called locusts—that emerged synchronously in the vicinity of Chicago in 1956 (brood XIII) numbered from well over 100,000 to about 1.5 million individuals per acre, as determined by Henry Dybas and D. Dwight Davis of the Chicago Natural History Museum. Since this brood extended over several hundred square miles, and there are 640 acres in a square mile, the total emergence consisted of many billions of cicadas. During the 17 years of their immature stage, these insects led solitary lives underground, sucking sap from the roots of trees.

True locusts, which are actually certain kinds of grasshoppers, are usually solitary and rather sluggish, but when they are crowded they enter a gregarious and highly active migratory phase. A swarm of migratory locusts that appeared in North Africa in the 1950s was estimated to include about 8 billion individuals and calculated to weigh 20,000 tons in the aggregate. Wherever the locusts landed, they ate almost everything that was green, leaving the land as barren as if it had been burned over.

Being part of a group can enhance an individual's fitness in a number of different ways, by helping it to satisfy, in one way or another, one or more of the three ecological imperatives that circumscribe the life of any living thing: it must eat and grow; it must avoid being eaten; and it must reproduce itself.

Group living can facilitate feeding. Just as a pack of wolves can subdue a large prey animal that one wolf could not overcome, perhaps a moose or a lone muskox, a column of army ants in South America or driver ants in Africa can overpower an insect or even a vertebrate that is too large for a single ant to handle. Some insects may even need to overpower the plants on which they feed, by joining forces to overcome special defenses that are effective against individual attackers. As John Byers noted, if enough bark beetles attack a tree at the same time, they can weaken it sufficiently to slacken the flow of resin that would otherwise drown them in the tunnels they excavate just beneath the bark.

A group of animals may also find food more quickly than can an individual. Birds, probably more so than other terrestrial animals, often benefit from foraging in flocks. A young bird may gain from associating with a more experienced flock member that is more adept at locating food, and the whole flock may benefit from the good luck of an individual that happens to make a lucky find. Honey bees, among the most highly social of all animal species, are the epitome of cooperative foraging. When a scout bee finds a new source of nectar or pollen, she returns to the hive and does a dance, known as the waggle dance, that communicates the distance and direction of the newly discovered patch of flowers to other foragers that cluster close to her and follow her as she dances. These followers, the new recruits, then proceed to the indicated food source, and, upon their return, they repeat the waggle dance, but only if they found the flowers to be productive. Thus poor food sources are ultimately abandoned and ever increasing numbers of workers are recruited to good sources. In this way, the attention of the colony is focused on a few productive patches of flowers.

Joining a group can help an individual avoid becoming a meal for a predator. Muskoxen, *Coelomera* beetles, termites, and owlflies put up active defenses, but there are other less obvious ways in which membership in a group tends to help individuals elude predators. For one, a predator that an individual does not notice might be discovered by another member of the group, who will then sound the alarm. Two pairs of eyes are better than one—as are two noses or two pairs of ears.

The mere presence of others—even if they are passive and neither give warnings nor participate in an active defense—decreases the probability that an individual will be taken by a predator. In effect, the individual gets lost in the crowd. Some insects, such as periodical cicadas and migratory locusts, are so numerous that the resident predators can eat only a small fraction of them. The appetites of the predators are thus saturated, so that there is only a small probability that any given individual will be eaten. A member of a group can benefit even if its companions are not numerous enough to satisfy the predators' appetites. For example, a lone individual who encounters a hungry predator too small to eat more than one prey animal is almost certain to be captured and eaten, but an individual in a group with nine others stands only one chance in ten of being singled out by the predator. Of course this protection will be less effective if a group is easier to discover than a lone individual.

According to W. D. Hamilton's concept of the "selfish herd," the members of some groups try to save themselves from predators by hiding among their fellows. It's every individual for itself. Sober and Wilson say that this is to be expected if *individual selection* is the only evolutionary force involved: "individuals will attempt to position themselves so that other individuals are between themselves and the predator." An individual is safest when it is in the midst of the group, as John Hudleston noted when he pointed out that small birds attacked mainly stragglers at the fringe of a band of marching desert locusts. But, Sober and Wilson argue, if *group selection* (or *kin selection*) is the only operative evolutionary force, there will be sentries or guards to protect the group, such as nasute termites, the soldier caste of some ants, or, as Jack Hailman and his coauthors observed, the sentinel system of Florida scrub jays, in which the members of a family group take turns watching out for predators.

❀ ❀ ❀   An animal that reproduces sexually must at some time during its life find a mate. Many solitary insects can find members of the opposite sex only by means of long-range visual, sound, or chemical signals. A lone female firefly perched on a blade of grass attracts a male by intermittently flashing in a code characteristic of her species. A male mole cricket chirps his species-specific song just inside the entrance to his burrow. Shortly before dawn or shortly after dusk a female cecropia, the largest of North America's wild silkworm moths, releases a sex-attractant pheromone, an odor peculiar to cecropia, that drifts downwind and may attract a distant male.

Some insects form groups to amplify signals that attract potential mates. Within their huge aggregations, which may number in the billions of individuals, periodical cicadas form smaller aggregations, numbering in the hundreds or thousands—choruses of males that sing in unison and attract both males and females. Fireflies in Indochina and other parts of Asia attract both males and females by forming huge aggregations that at night flash in synchrony, often so brightly that they light up an entire tree. Massing together benefits the bark beetles that burrow in conifers, not only because they combine forces to overcome the defenses of the tree but also because they produce a stronger and more far-reaching pheromonal signal than can a lone beetle, a signal that attracts potential mates of both sexes.

For some gregarious insects, such as mayflies and migratory locusts, long-distance communication is unnecessary. Members of the opposite sex are not difficult to find; they will be nearby and plentiful. But an eager suitor must still determine if a potential mate is actually a member of its own species and a member of the opposite sex. This is not as simple a problem as it is for people. Even an insect that is gregarious as an adult may never have seen another adult of its own species until it itself becomes an adult. Insects solve this problem by means of species-specific recognition signals. The ability to produce and recognize these signals is inborn, programmed by the genes. These signals are sometimes visual or auditory but are most often chemical, close-range pheromones that are perceived by taste or smell.

❁ ❁ ❁   The following chapters are arranged according to the basic needs served by group living: defense against enemies, coping with the weather, finding and subduing food, and meeting members of the opposite sex. In all cases that I know, group living provides more than one benefit. For example, the mass emergences of mayflies not only bring the sexes together but also lessen the threat to any one individual from predators. Although the next three chapters touch on several topics, they are primarily concerned with group defenses against predators, such as the collective advertisement of their inedibility by various insect species, ranging from relatively small groups of red and black overwintering ladybird beetles to winter concentrations in Mexico of tens of millions of orange and black monarch butterflies. The bright colors of these insects, amplified by their "cheek by jowl" abundance, warn birds and other predators of their toxicity. In the next two chapters I discuss, among other issues, some of the ways in which group living makes it possible to ameliorate adverse weather conditions. Then I turn to how insects and some other animals synchronize their activities to coincide with food resources and how they gang up to subdue their food if it fights back, be it animal or vegetable. After that I move on to the vast swarms of billions of migrating locusts, the ultimate in group behavior in the search for new sources of food, and then describe some of the ways in which people have sought to deal with plagues of locusts and other insects. Next I discuss group activities to find members of the opposite sex, activities that range from frequenting the insectan equivalent of singles bars to massing together to amplify visual or auditory signals that attract potential mates. In the final chapter I consider the ultimate in adaptive togetherness, the formation of cooperating groups that include two or more species.

Bunches of Beetles

*Ladybird beetles just after waking
from their winter sleep*

❀ ❀ ❀   It is a warm day in early April—one of those days when it is great to be alive because the air temperature is just right, the sky is a shade of blue that talks directly to the soul, and life is burgeoning all around you as plants and creatures awaken from their long winter sleep. This is the time of year when the woodland wildflowers flourish on the forest floor, their photosynthetic processes driven by the sunlight that shines through the crowns of the trees, whose branches are still bare, with their leaf buds just beginning to swell. In another month the buds will have burst and the leaves will form a dense canopy that casts a shade too deep for the spring wildflowers to tolerate. But today the forest floor is still sunny; the foliage of trout lilies, wake-robins, and May apples has poked up through the layers of fallen leaves that carpet the ground, a promise of blossoms to come; here and there a bloodroot or a clump of Dutchman's-breeches already flaunts its blossoms in hope of attracting pollinating insects; spring beauties are everywhere, their small pink blossoms visited by flies and little brown bees.

As I walk through this woodland on a bluff above the Sangamon River in central Illinois, I am aware of the insects around me, not yet as abundant as they will be later in the year, but already numerous enough to satisfy the appetites of a few early spring migrant birds, among them a fox sparrow that kicks aside fallen leaves to expose seeds and crawling insects and an eastern phoebe that darts from its perch to snap flying insects from the air. On this day I have the good fortune to see a mourning cloak butterfly as it flits among the trees. Mourning cloaks, usually the first butterflies to be seen in spring, spend the winter in lonely isolation in a hollow tree or some other protected nook. The English, recognizing the somber beauty of this insect, which occurs naturally both in the Old and New Worlds, have named it the Camberwell beauty.

A patch of red off to one side of the trail catches my eye. It is quite large, covering more than a square foot of ground and extending several inches up the trunk of a tree. When I move closer, I see that it is a densely packed aggregation of thousands of little ladybird beetles that are just coming out from under the layers of fallen leaves where they spent the winter. Some of them, ravenous after their long winter fast, have crawled onto the blossoms of nearby spring beauties, where they are eating pollen in the absence of their usual diet of aphids and other soft-bodied insects.

These ladybird beetles, like the mourning cloak and the other insects that flew or crawled in the woodland that day, had just awakened from the hibernation-like state that entomologists call diapause. The ability to diapause is crucial to most temperate zone insects—and even to some tropical insects. An insect that is in diapause can survive a long winter in the north or a dry season in the tropics that it could not otherwise live through. No doubt these ladybirds, like most other insects, had been triggered to go into this dormant state the previous year by the short days of late August and September, the only consistently reliable fore-warning of the approaching winter. (Insects may diapause in any one of their life stages but usually all individuals of a species diapause in the same life stage.) Diapausing insects stop developing: In insects with *complete metamorphosis,* such as butterflies, beetles, and bees, embryos do not mature and eggs do not hatch; larvae do not feed, grow, or molt to become pupae; pupae do not metamorphose to become adults; and adults neither mate nor lay eggs. It is much the same with insects with *gradual metamorphosis,* such as grasshoppers, dragonflies, and true bugs (members of the order Hemiptera), which have no pupal stage: eggs do not hatch; the immatures, known as nymphs, do not metamorphose to the adult stage; and adults neither mate nor lay eggs. Insects that are in diapause can withstand subfreezing temperatures without damage to their bodies, and most of them—but not all—are physically inactive. Their metabolic rate falls from one tenth to as little as one twentieth of the usual rate, allowing them to survive on their stored body fat from ten to twenty times longer than they could if they were not in diapause.

Diapause must be terminated at the appropriate point in the spring to give the insect enough time to attain the adult stage if it is an immature, and for it, and perhaps also for one or more generations of its descendants, to reproduce before the onset of winter. There are several mechanisms for the termination of diapause, but many insects—including pupae of the handsome cecropia moth and others of the giant silkworm moths of North America—are reactivated by the warmth of spring. But this mechanism presents a problem: the insect will be doomed by winter weather if it prematurely terminates diapause in response to warm temperatures in autumn or to an unseasonable period of warm weather in winter. Cecropia and many other insects, probably including ladybirds, avoid this deadly trap by responding to a period of warm weather only *after* they have been exposed to a sufficient duration of cold weather. They keep tabs on the amount of *coldness* they have experienced—a physiological process but not unlike a meteorologist's keeping track of heating degree days in winter—and do not terminate the diapause state in response to warmth until they have experienced enough coldness to tell them that winter has passed. (You can get a cecropia moth to emerge indoors from a cocoon collected before the onset of winter by holding the cocoon in a refrigerator for 10 weeks as a substitute for the natural period of chilling.) This reactivating effect of cold weather distinguishes the diapause of insects from the hibernation of vertebrates. Hibernating vertebrates do not require previous exposure to cold to become reactivated by warmth. Bats and groundhogs, for example, may become briefly active during a spell of warm weather in winter and then resume hibernating.

The diapause-terminating mechanism has another important function: synchronizing the reactivation of the individuals of a species. This function is most apparent in gregarious insects, such as our ladybirds, but is important to all insects. Even solitary insects benefit from becoming active in synchrony with the other members of their species. For one thing, the probability that an individual will find a mate is greatest if it emerges at the same time as the other members of its species.

❁ ❁ ❁   The ladybirds, also known as ladybugs, are a fairly large family, about 475 species in North America and about 4,200 worldwide. Only 2 of the North American species are plant eaters: a species that eats squash leaves but is only ocasionally numerous enough to be a garden pest and the infamous Mexican bean beetle, all too often a pernicious defoliator of garden beans and now in the process of adapting to the nonnative soybean and threatening to become a widespread pest of this important crop. All of the other North American ladybirds, both as larvae and adults, are predators that eat scale insects, mealybugs, aphids, and sometimes other soft-bodied insects.

Adult ladybird beetles are familiar to almost everyone, easily recognized as members of the ladybird family by their distinctive appearance. Their convex and rather rotund bodies are nearly hemispherical, and the great majority of them are red, orange, or yellow with a few black spots or black with a few red, yellow, or white spots. The larvae, which look a bit like miniature alligators, are more or less flattened and carrot-shaped; usually clothed with spines or wart-like bumps; and usually brightly colored, often with a dark ground color and patches of red, orange, yellow, cream, or white.

Ladybirds live exposed on plants, usually on the foliage, in all four stages of their life cycle—as eggs, larvae, pupae, and adults. The eggs, which are yellow or orange, are laid in closely packed clusters of one or two dozen—usually on a leaf and often in plain view on its upper side. The active and agile larvae can be seen running about on foliage, usually near or within a colony of the aphids or scale insects that are their prey. A larva that is ready to molt to the pupal stage fastens itself to a leaf or stem. Its skin splits down the back and the pupa remains more or less within the molted larval skin, or the skin peels back to become a tight little wad at the tail end of the pupa. During the warm seasons, adult ladybird beetles, which are not then gregarious, live on plants, where they prey on small insects, mate, and lay their eggs. They spend the winter in large groups as I have described, hidden under fallen leaves or other debris on the ground.

Ladybird beetles have a well-deserved reputation for being highly beneficial. There is no doubt that they, along with other predators and parasites, prevent populations of plant feeders from increasing to destructive levels. This has been unintentionally demonstrated in two ways. First, the use of insecticides on crops or in orchards has often caused huge increases of formerly uncommon native plant-eating insects by destroying their parasites and predators, including ladybirds. What happens is that relatively rare plant-feeding insects that are immune to an insecticide become abundant enough to be economically significant pests if the insects that eat them, including ladybirds, are killed because they are not immune to the same insecticide. Second, the

importance of ladybirds in suppressing the populations of other insects has been demonstrated by the results of a number of pest-control
programs that amounted to unintentional large-scale experiments: programs that imported foreign ladybirds into North America to control
plant-feeding aphids or scale insects that were accidentally brought
here from abroad and that ran wild because North American predators did not keep their populations in check. These foreign pests were
brought under control by introducing ladybirds from the native range
of the pest. This form of insect control, known as biological control, has
often been employed to control other insects and weeds by introducing
various pathogens, predators, or parasites of pest insects.

The most famous of the foreign ladybirds used as a biological control
is the vedalia, introduced into California from Australia in 1888 to control an exceptionally destructive insect, the cottony cushion scale. The
scale had been unintentionally brought in from Australia and was about
to destroy California's infant citrus industry. At that time the available
insecticides were stomach poisons—poisons that are sprayed onto the
surface of a plant and do not kill unless they are ingested. They do not
kill insects that, like aphids, true bugs, and the cottony cushion scale,
pierce the plant with their beaks and suck sap from below the surface.
From an ecological point of view, the lack of an appropriate insecticide
was a good thing, because entomologists had to look for controls other
than a quick but temporary fix from an insecticide.

Charles V. Riley, first director of the Division of Entomology of the
U.S. Department of Agriculture, as Richard Doutt has recounted, knew
that the cottony cushion scale never becomes abundant enough to do
major damage in its native Australia, although in California it was then
astronomically abundant and very destructive. He reasoned that this
scale was controlled in Australia by a natural enemy that was absent
from California. He was absolutely right. After the vedalia was imported and had become established in California, cottony cushion scale
populations plummeted, and to this day they are held at low, non-
destructive levels by correspondingly small populations of vedalias
with which they coexist. The vedalias failed only once, but it wasn't
their fault. In the 1940s DDT was sprayed on citrus trees to control insects other than cottony cushion scales. DDT killed vedalias but not the

scales. The scale population burgeoned and once again was destructive. But when another more ecologically sound insecticide that did not kill ladybirds was substituted for DDT, the vedalias recovered and the scale population once again plummeted and has remained low ever since.

Ladybird beetles are among the few insects that are admired and liked by people. People who would be disturbed if a fly or a grasshopper landed on their body let ladybird beetles crawl freely on their arms and hands. English-speaking children let these little creatures sit on their hands as they chant a nursery rhyme:

> Ladybird, ladybird, fly away home,
> Your house is on fire, and your children will burn.

Quite a few languages have endearing names for ladybird beetles, often derived from religious terms. Our English name is a shortened form of Our Lady's bird. In German the ladybird is called *Marienkäfer,* Mary's beetle. The French know it as *bête á Bon Dieu,* the creature of God. The Dutch call it by a name with about the same meaning, *Lieveheersbeestje,* the dear Lord's little creature. In Hebrew the ladybird is *parat Moshe Rabbenu,* the creature of Rabbi Moses, and in Greek these beetles are called *paschalitsa,* the little ones of Easter.

⚙ ⚙ ⚙   When I returned to the woodland along the Sangamon about a week later, the aggregation of ladybirds was gone. Overwintered ladybirds must soon disperse to lead largely solitary lives as they feed and reproduce. It could not be otherwise. Even a small group of ladybirds would soon outstrip its food supply even if it lived within a colony of aphids that included dozens or even hundreds of individuals. Ladybirds can, however, afford to become gregarious in winter, because then they live on the fat stored in their bodies and don't have to feed. But why do these beetles gather together in such huge overwintering aggregations? A better way to phrase this question is to ask how they benefit from aggregating. There may be several benefits. Crowding together could conserve water by slowing down the dissipation of moisture lost from the bodies of the beetles, as has been demonstrated with handsome fungus beetles, to be discussed in a later chapter. Another potential benefit, and possibly the major one, is that aggregating tends to protect them against predators, as you will see in the next chapter and has been well established for monarch butterflies.

❀ ❀ ❀ People have taken advantage of the gregariousness of ladybird beetles in attempts, unfortunately futile, to use them as control agents by collecting them in winter and then in spring releasing huge numbers of them where aphids or other natural ladybird prey have become or might become pests.

In the winter of 1910, two men dug down through the deep snow at a marked spot on a slope high in the Sierra Nevada of California. When they reached the surface of the ground, they brushed aside pine needles and other debris to reveal a large aggregation of ladybird beetles. During the previous autumn, these ladybirds had flown up from the valleys below to spend the winter diapausing high in the mountains. (The ladybirds that I saw in Illinois and the ones that overwinter in the Sierra Nevada of California are two different species, *Coleomegilla maculata* in Illinois and *Hippodamia convergens* in California, but both are, like many ladybirds, colored red and black.) They scooped up the beetles, a mass larger than two of a man's fists, passed them through a screen to separate them from debris, and put them in a cloth bag. That day they would locate several more such aggregations and would collect all together about 125 pounds of ladybirds, well over 2,800,000 of them. Sometimes the beetles were carried by a mule, but under some circumstances it was not practical to bring along a pack animal. As E. K. Carnes, then Superintendent of the California State Insectary, wrote:

> The insects are bulky, relatively [*sic*] to the weight, and so a mule can not very well carry more than 125 pounds of them. More often, however, the colonies [aggregations] are so inaccessible, and so far off from the mountain trails, that they are carried by the field men themselves into camp, slung in sacks over their shoulders. The trails in the high Sierra are usually so steep, and the footing so precarious, that riding is dangerous, and the men usually prefer to "hoof it." If a saddle mule be taken along on such an expedition it is allowed to go first, and the field men have the animals pull them up the steep grades, by hanging on to their tails.

This aggregation of ladybird beetles in the Sierra Nevada had first been located late the previous autumn, before the snow had fallen and while the beetles, although largely covered by pine needles, were still visible to a trained eye. Beginning about the first of November, scouts had gone up into the mountains to locate clusters of the hibernating beetles before they were covered by snow. They were most often found on sunny, well-drained slopes, usually near running water. As Carnes reported:

> Whenever a colony is located, we dig into the pine-needles, moss, leaves, etc., and through past experience we are enabled to estimate about how many pounds the said colony will yield. A little map is then roughly drawn on a card, a tree is blazed marking this spot, and the colony numbered with a notation on the back of the card, giving conditions, probable amount obtainable, together with any information the collector thinks would be useful during the winter when [the ground is covered with snow and] the real work of capturing the colony is to take place.

As Carnes reported, several tons of the ladybirds were collected each winter. He did not specify just how many tons, but the total number of beetles collected was immense. A single ton of beetles consists of 45,344,000 individuals.

About once a week a pack train of mules came up the mountain trail to the collectors' camp to bring supplies and to haul the sacks of beetles that had accumulated to a small packing house at the closest railroad station, which was still in the mountains but about 12 miles away. Here the beetles were further cleaned and packed in small wooden crates that held about 33,000 beetles each. They were then held in the cold mountain packing house as long as possible, but before warm weather arrived they were moved by train to a cold-storage facility in Sacramento, where they were held at a temperature just below 40° Fahrenheit. This cold treatment, as Carnes explained, kept the beetles in a state of "artificial hibernation" (diapause) that could be maintained for several months if necessary.

Ultimately, the beetles were shipped free of charge by the horse-drawn wagons of the Wells Fargo Express Company to farmers and orchardists who believed that their crop of cantaloupes or other fruits was threatened by aphids. A grower was entitled to a consignment of one crate of beetles for each 10 acres of threatened crop. The hope was that adult beetles that were released in the field would begin the job of destroying the pestiferous aphids, but that they would also lay eggs and that their larvae, which also eat aphids, would carry on.

Numerous testimonials praising the work of the ladybirds, which were provided at the expense of the state of California, were sent to Carnes and to the State Commissioner of Horticulture in Sacramento. A letter received in June of 1911 said, "Upon my arrival at Brawley to handle about . . . two thousand acres of cantaloupes, I found the vines in most excellent condition, free from all aphids. Inquiring about the care that had been given to them, I was pleased to learn that the credit is given to the ladybirds."

This anecdotal testimonial and the others received were not, however, based on systematic observations. Unfortunately, systematic observations of released ladybirds made by W. M. Davidson in 1919 and 1924, by C. M. Packard and Roy E. Campbell in 1926, and by Philip Garman in 1936 showed that only a few days after their release almost none of the beetles could be found near the release point, that they had dispersed widely to areas as much as 6 miles away. It thus appears that any perceived control of aphids is due not to the beetles that were released but, rather, to other factors, perhaps native beetles, often of the same species, that came to the crop field on their own.

In the article that he published in 1924, W. M. Davidson reported the results of a mark and recapture experiment with artificially released ladybirds that he had done in March of 1919 in the Imperial Valley of California. He released about 610,000 living beetles that had been sprayed with gold-colored aluminum paint. The great majority of the beetles flew off almost immediately. Shortly after the release, the area was repeatedly searched for marked beetles. On the third day after the release only 6 were found within a mile of the release point; 1 was found at something over a mile away; and 2 were found almost 5 miles away.

It is no surprise that the beetles dispersed soon after they were released. They had been kept cold and in diapause until the time of release. When they awakened in the warmth of the release site, their instincts told them that they had to disperse or die. As far as the beetles knew, they were still high up in the Sierra Nevada. If they could think and if we could read their minds, we would probably hear something like this: "I had better fly a long distance away in the hope of escaping from this horde of competitors and getting down to the valley where aphids are abundant."

As recently as 1998, J. J. Obrycki and T. J. Kring wrote that no evidence to contradict these early studies on the dispersal of ladybirds has appeared in the intervening 60 or so years. They point out not only that the mass release of such wild-caught ladybirds is futile, but that it probably has undesirable consequences. First, the removal of vast numbers of these aphid-eating beetles from the mountains depletes their population and may thereby have an adverse effect on crops grown in the valleys below. Second, transporting these beetles to all parts of the country may disseminate diseases that are injurious to native local populations of ladybirds of the same or other species.

Despite evidence to the contrary, some people still try to control pest insects by making inundative releases of large numbers of diapausing ladybirds in their gardens. Freelance collectors still earn money by gathering these beetles in the mountains of the west, and, as reported in 1964 by Paul DeBach and Ken Hagen, even at that late date some dealers sold as many as 10,000 gallons of these ladybirds each year. That is almost three quarters of a billion beetles. These ladybirds are still for sale today. They are widely advertised in gardening magazines and in the catalogues of garden supply houses. I have no idea how many of them are sold these days, but with the current interest in organic gardening and natural pest-control methods they are likely to be even more popular than they were in 1964. As I write, I have before me a garden-supply catalogue that offers ladybirds for sale at about $50 per quart, about 18,000 beetles. No mention is made of the fact that the beetles are likely to disperse widely almost as soon as they are released. These beetles are obviously still in diapause because the catalogue specifies that they can be stored in a refrigerator for a long time.

❄ ❄ ❄ Ladybird beetles are not usually eaten by birds or other predators because they are toxic and unpalatable. Many other insects are similarly protected by venomous stings or by toxins that they contain in their bodies, sometimes toxins that they obtain from their food plants. The great majority of these protected insects, like ladybird beetles, warn predators of their noxiousness by means of easily recognized signals such as conspicuous coloration or distinctive odors. There is ample evidence that these warning signals are heeded by predators and often save the lives of insects. As you will read in the following chapter, coming together in groups can amplify and thus enhance the effectiveness of such warning signals.

Warding Off Predators

*Sleeping zebra butterflies clinging
to the twigs of a small bush*

❊ ❊ ❊   Most animals, ranging from insects to mammals, have behaviors, characteristics, or defensive weapons that protect them to some degree from predators. Some try not to be noticed. Most small mammals and many insects are camouflaged and thus difficult to spot because they blend in with the background; others, mostly insects, resemble some object, often a bird dropping or a stubby twig, that is of no interest to predators. But many insects have defensive weapons that deter predators. Some insects, ladybirds and monarch butterflies among them, contain, ooze out, or spray toxins that sicken predators that eat or try to eat them. Others, including wasps, ants, and bees, have venom-injecting stingers, and certain caterpillars have sharp, fragile hairs that break off in the skin of an attacker and release a venom that stings like a nettle. These caterpillars are said to be urticating, from the Latin *urtica*, a nettle.

Charles Darwin's encounter with a chemically armed beetle, probably some species other than a ladybird, is dramatic and convincing testimony to the effectiveness of the defensive secretions of insects. In his autobiography—edited by his granddaughter Nora Barlow—Darwin, an enthusiastic collector of beetles, recounted an experience that he had while trying to capture some particularly interesting specimens: "one day, on tearing off some old bark, I saw two rare beetles & seized one in each hand; then I saw a third & new kind, which I could not bear to lose, so that I popped the one which I held in my right hand into my mouth. Alas it ejected some intensely acrid fluid which burnt my tongue so that I was forced to spit the beetle out, which was lost, as well as the third one."

When they are disturbed, members of the ladybird family discharge a bitter, amber-colored fluid that oozes from pores near the "knee joints," the articulations between the two longest segments of the leg. This fluid is blood laced with astringent and odoriferous chemicals that are synthesized by the beetles. By 1903 it was known that this secretion is toxic to vertebrates, and in 1960 J. F. D. Frazer and Miriam Rothschild re-

ported that all of the many birds and small mammals that they tested refused to eat ladybird beetles. As a matter of fact, ladybirds were at the top of the list when Frazer and Rothschild ranked 19 toxic insects for their repulsiveness to small vertebrate predators.

The red and black coloration of ladybird beetles makes them highly conspicuous. Being so easily spotted would be decidedly disadvantageous for a palatable insect that birds, mice, or other predators like to eat. But it is an advantage to a ladybird beetle or any other well-defended animal, an obvious, easily understood, and readily learned warning to would-be predators of its noxiousness. Skunks, whose potent chemical defense is well known, are warningly colored in black and white, the only small mammals of North America that are not camouflaged. Although some predators have an innate—genetically programmed—understanding of such warnings, most of them must learn their meaning by experience. Consequently, if an inexperienced predator is to learn to shun ladybirds, monarch butterflies, or the many other warningly colored and noxious insects, it must at some time attack and perhaps kill at least one of them—possibly even more of them if the toxin content of an individual is less than an effective dose, or if the predator's memory is poor and must be refreshed from time to time.

An individual ladybird can minimize the probability that it will become an object lesson for a predator by joining an aggregation, by getting lost in the crowd and surviving at the expense of the few other ladybirds that are sampled by the predator. An aggregation is also a more conspicuous warning than a lone individual. In winter, when they are hidden under debris, the collective odor of an aggregation of ladybirds may warn away a white-footed mouse or a shrew that might disturb them as it rummages through fallen leaves for a meal. When the ladybirds first leave their hiding place in spring, their massive aggregations are visible from a distance and may warn away birds and other predators before they even come close.

❁ ❁ ❁   Animals that advertise their noxiousness with bright colors, odors, or other signals are said to be aposematic, giving off warning signals, a word that is often incorrectly understood as meaning no more than "warningly colored." The warning signal is often the animal's color pattern, but it can also be the buzzing sound of a rattlesnake, the angry buzz of a wasp, or the acrid odor of a toxic lubber grasshopper. The yellow and black of a stinging wasp or bee, the black and white of a chemically armed skunk, or the red, yellow, and black of a venomous coral snake demand the attention of vertebrates. Even people use these colors as warnings. Along our roads, stop signs and stoplights are red, and signs that warn of danger ahead are yellow and black.

There have been many experimental demonstrations of the effectiveness of insects' warning signals, but an exceptionally interesting and convincing one was done by Thomas Boyden in Panama. The predators in his experiments were wild, free-ranging, ground-dwelling *Ameiva* lizards, insect-eating animals large enough to grab and consume the butterflies that Boyden presented to them as potential prey. He used two species of butterflies: the palatable and not warningly colored *Anartia fatima;* and the impressively warningly colored black, yellow, and red *Heliconius erato,* already known to be repulsive to birds from experiments done on Trinidad by Lincoln Brower and his colleagues. (*Heliconius* butterflies, as Kevin Spencer has noted, are noxious because they contain cyanide compounds that they sequester from the passionflower plants on which they feed.) Boyden cast living butterflies to the lizards using a spinning reel, fiberglass rod, and 10-pound test nylon line. All the butterflies, Boyden wrote, "were tied with lightweight green thread by a loop which extended around the wings at their base, in such a manner as to allow the butterfly to fly freely in a normal fashion." The free end of this leash was tied to a swivel at the end of the nylon line and "a single shot sinker was used as a weight for casting the line." The lizards readily attacked these butterflies as they fluttered at the ends of their leashes, and they seemed to have no difficulty in catching and eating them.

Boyden's results leave no doubt that the lizards found the *Heliconius* butterflies to be noxious, and that they often heeded their warning coloration before attacking them. When 19 palatable and not warningly colored *Anartia* were cast to the lizards, the lizards attacked 95 percent of them, killed 89 percent, and ate 84 percent. But when they were offered noxious and warningly colored *Heliconius,* they attacked only 47

percent of them, killed only 22 percent, and ate none. As these data show, the protection given by the noxiousness and warning signals of the *Heliconius* is not perfect, but it did save 78 percent of them from death in the jaws of a lizard.

The coloration of the *Heliconius* butterflies obviously acted as a warning signal that, about half of the time, warded off potential attackers before they so much as touched the butterfly. This point was driven home by two additional experiments by Boyden in which *Anartia* and *Heliconius* with their visual warnings obliterated in two different ways were cast to the lizards. In one of these experiments, *Heliconius* butterflies with the scales, and thus the color, rubbed off their wings were attacked almost as frequently (90 percent of the time) as were intact, palatable, and not warningly colored *Anartia*. Although they were attacked at first, many of these *Heliconius* were subsequently rejected, presumably because of a noxious flavor or smell. Only 56 percent of them were killed and only 44 percent were actually eaten. By contrast, 100 percent of the *Anartia* were killed or eaten. In the other experiment, Boyden compared the attack rate on *Heliconius* that had the red and yellow areas of their largely black wings covered with black paint to the attack rate on control *Heliconius* with their warning colors intact. Black paint had been put on black areas of the wings of the control butterflies not to change their appearance but to assure that they were equally affected by handling and the presence of paint on the wings. *Heliconius* with their red and yellow colors still visible were attacked by the lizards only 50 percent of the time, but those with these colors blacked out were attacked 100 percent of the time. Nevertheless, both types were killed only 33 percent of the time and none were eaten.

A laboratory experiment done by Birgitta Sillén-Tullberg, a Swedish biologist, again demonstrated that warning signals are effective and that the combination of noxiousness plus warning signals gives more protection against predators than does noxiousness alone. She presented caged great tits, birds related to chickadees, with two forms of a noxious bug, the normal red, warningly colored form, and an equally noxious mutant gray form that is not warningly colored. The birds initially attacked both red and gray bugs, but soon learned to avoid both forms. During the learning period, however, the warningly colored bugs had a higher survival rate than the mutants, because the birds were more reluctant to attack them, learned to avoid them earlier on, and less often killed them during an attack.

⚙ ⚙ ⚙ Of course groups—including groups of aposematic creatures—vary in size, cohesiveness, permanence, and the activities that they perform. Some noxious animals, such as certain grasshoppers, bugs, beetle larvae, caterpillars, and sawfly larvae, usually form only relatively small groups, and then, as a rule, mainly when they are feeding on a plant. Some insects that are not gregarious during the day, including certain toxic butterflies and stinging wasps, form more or less tightly packed but usually small sleeping aggregations at night. Others, among them monarchs and ladybird beetles, come together only in autumn, as groups of diapausing individuals that stay together throughout the winter. Ladybird beetles, as we saw, form overwintering groups that often consist of thousands of individuals. Monarch butterflies, as you will read in the next chapter, form overwintering aggregations of tens of millions of individuals. Some of the social wasps that make uncovered paper combs, which you may have seen hanging from the eaves of a house, form smaller overwintering groups that may include only a few dozen fertilized queens. (Wasps make paper by combining wood fibers with their saliva. It is said that the Chinese, the first people to make paper, learned the process by watching wasps.) A few animals, notably certain fish that form schools, may occur in very large groups and remain together and active for long periods of time, sometimes even more or less permanently, such as certain marine catfish, discussed below, that have venomous spines and form tightly packed clusters to warn off predators.

An aposematic grasshopper that occurs in Argentina forms large feeding aggregations during its nymphal stage. William Henry Hudson, best known for his novel *Green Mansions*, described the habits of this grasshopper in *The Naturalist in La Plata*, first published in 1892:

> The young are intensely black, like grasshoppers cut out of jet or ebony, and gregarious in habit, living in bands of forty or fifty to three or four hundred; and so little shy, that they may sometimes be taken up by handfuls before they begin to scatter in alarm. Their

gregarious habits and blackness—of all hues in nature the most obvious to the sight—would alone be enough to make them the most conspicuous of insects; but they have still other habits which appear as if specially designed to bring them more prominently into notice. Thus, they all keep so close together at all times as to have their bodies actually touching, and when traveling, move so slowly that the laziest snail might easily overtake [them] . . . They often select an exposed weed to feed on, clustering together on its summit above the surrounding verdure, an exceedingly conspicuous object to every eye in the neighborhood.

Hudson went on to say that he had never seen a bird prey on these grasshoppers, although he watched some bands of the grasshoppers off and on for several days "in places where the trees overhead were the resort of Icterine and tyrant birds, Guira cuckoos, and other species, all great hunters after grasshoppers." The ebony-colored grasshoppers were safe, presumably because they taste nasty, despite the fact that they are conspicuous and could be easily captured by any predator that comes along.

🔹 🔹 🔹   The chemical defenses of some insects—among them the monarch and probably the grasshopper described by Hudson—are contained within the body, and, consequently, do not act until the insect is eaten by a predator or at least held in its mouth. Such a defense may not protect the unfortunate individual that is attacked, but its sacrifice is not necessarily in vain, because, as you already know, it will probably teach the predator not to attack other members of the victim's species—in gregarious species, often its own brothers and sisters—that make up the defensive group. Even if the victim dies, if its sacrifice discourages the predator from making further attacks on the group, it will have enhanced its own inclusive fitness by protecting kin in which its genes will survive.

But some insects, among them ladybird beetles, stinging wasps, sawfly larvae known as spitfires, and urticating caterpillars, can ward off a predator before they are eaten or seriously injured. Obviously a toad that is stung in the mouth by a bumble bee is likely to release its offensive prey unharmed. A large nongregarious red and black assassin bug of East Africa "spits" when it is disturbed, most likely to ward off insect-eating monkeys and other vertebrates. J. S. Edwards reported that the bug's saliva, which can be squirted as far as 12 inches from its beak, is extremely irritating to the membranes of the eye or nose. The venom that is released when the sharply pointed hairs or spines of urticating caterpillars pierce the skin and break off is always highly irritating and may sometimes, as is the case with the caterpillars of South American flannel moths, cause severe pain, swelling, fever, and even paralysis. As Walter Linsenmaier reported, these menacing flannel moth caterpillars are known as *bizos de fuero* (fire beasts) in Brazil and as *iso jagua* (jaguar worms) in Paraguay. Some urticating caterpillars are solitary, as are those of the flannel moth in South America and the io moth in North America. Others are highly gregarious. The spitfires, sawfly larvae that live in Australia, cluster together in dense aggregations while they feed on the leaves of eucalyptus trees. As reported in an exceptionally interesting article by P. A. Morrow and two coauthors, they store a greasy fluid, apparently derived from the oils of eucalyptus leaves, in their guts. If threatened by a predator, the colorful larvae rear up in unison, and each one regurgitates a large drop of the oil, which deters attacks from ants, birds, and mice.

The infamous urticating pine processionary caterpillars of Europe, closely observed by H. H. Brindley, are also gregarious. A group of them, as many as 300 strong, builds a communal tent of silk on a branch of a pine tree. During much of the day they bask in the sun as they cling to the outside of the tent, but shortly after dark they move away to feed, forming a procession as they crawl in single file nose to tail. After having fed, they return to their tent an hour or two after midnight and stay inside it until they come out to bask at midmorning. These caterpillars are brightly aposematic: dark-bodied, sparsely clothed with long white hairs, and marked with bright red spots that are set with thousands of tiny and easily broken urticating hairs. In daylight, their appearance is obviously a warning to would-be predators. But do they give warning to nocturnal predators that might attack them as they feed at night? Perhaps they do by giving off some nonvisual warning signal, such as a distinctive odor.

Throughout much of the tropical and subtropical New World, from

the southern United States southward, a huge (as much as 5 inches long), gregarious, and flagrantly conspicuous hawk moth caterpillar (*Pseudosphinx tetrio*) attacks frangipani trees, popular ornamentals in warm climates. During the day, these caterpillars cluster in conspicuous groups on bare branches, and in the dark of night they spread out to devour leaves. They are ringed with jet black and bright yellow; their legs, heads, and tail ends are a brilliant red-orange; and they have long black tails that they often lash from side to side. If they are touched, they thrash back and forth and, unusually for a caterpillar, bite viciously. These spectacularly conspicuous creatures are obviously trying to tell us something. The message is probably that they are not good to eat because they are poisonous, most likely because they sequester in their bodies toxins that are in the frangipani leaves they eat. In other words, they are probably truly aposematic, giving warning of a real threat to a predator that might attack them. This view was put forth in 1920 by the Reverend A. Miles Moss, the British chaplain of Pará, Brazil, and a dedicated student of the hawk moths, in an article, beautifully illustrated in color, that was published in *Novitates Zoologicae,* the official journal of Lord Walter Rothschild's private museum at Tring in England. Moss enumerated the many enemies of unprotected hawk moths, including reptiles, birds, and mammals, but noted, as have other observers, that predators do not attack frangipani caterpillars.

An opposing view was offered by Daniel Janzen. He suggested that frangipani caterpillars are not poisonous, but gain some protection from predators by bluffing, by mimicking the colors of venomous coral snakes. Janzen is probably wrong about the frangipani hawk moth caterpillar. (Since the palette of warning colors is limited, it would not be unusual for different noxious creatures, such as caterpillars and snakes, to evolve similar warning signals.) But there are many other well-documented instances of such protective mimetic resemblances. Unusual among vertebrates but common among insects, this kind of mimicry is known as Batesian mimicry, named for Henry W. Bates, who described the first recognized case in 1862, an edible butterfly of the Amazon Valley that so closely resembles an unrelated, inedible butterfly that the two are difficult to distinguish even when they are held in the hand. There are many other cases of Batesian mimicry: harmless flies, moths, beetles, and even grasshoppers that closely resemble stinging wasps or bees; palatable butterflies that resemble toxic butterflies; innocuous spiders, bugs, and beetles that look like stinging ants; edible cockroaches that closely mimic unpalatable ladybird beetles or leaf beetles.

⚙ ⚙ ⚙ Early on a July morning in 1913, Phil and Nellie Rau discovered a sleeping aggregation of the steely-blue, spider-hunting wasp *Chalybion caerulium* on the ceiling of an open cowshed in Kansas. During the day, these wasps are solitary. Lone males search for lone females, and females hunt for spiders with which to stock the nests where their larvae will develop. (A *Chalybion* female appropriates for herself the mud nests of other wasps, removing the contents placed in the nest by its rightful owner and then filling it with spiders that she paralyzed with a sting. She lays an egg on the last spider to go in and seals the nest with mud.)

I have no doubt that the wasps observed by Phil and Nellie Rau are protected against predators either by the same sting that they use to subdue spiders, although it seems mild to humans, or by the presence of some noxious substance in their bodies. Their shiny blue coloration and their habit of running about erratically and nervously as they flick their dark wings make them very conspicuous and are surely intended as warnings. A telling indication of their noxiousness is that they are closely mimicked in both appearance and behavior by a totally harmless and edible hover fly, *Xylotomima chalybea*. There is thus little doubt that *Chalybion's* sleeping aggregations are a defense against predators—perhaps because they ward off all predators with some massed warning signal, or perhaps because an individual predator is turned away from the group by the sacrifice of only one or a few of their number, a sacrifice that informs the predator that it has encountered noxious and inedible insects, and thus dissuades it from eating any more.

The Raus found that these wasps began to arrive at their sleeping place at dusk, that they were fast asleep by dark, and that about an hour after dawn they began to depart, one by one. As many as a hundred or more crowded into a small space of less than 1 square foot. Night after night, the *Chalybions* assembled in the exact same spot on the ceiling of the cowshed, and even after Phil and Nellie had been away for 6

weeks, they found a large congregation of these wasps in the very same spot. The wasps may have marked their sleeping place with an odorous pheromone.

Many aposematic insects other than *Chalybion* are solitary during the day but form sleeping aggregations at night, among them many other wasps and various bees, butterflies, and grasshoppers. An early account of butterflies sleeping in groups appeared in the *Atlantic Monthly* in 1918, written by William Beebe, a famous naturalist of the early twentieth century and best known as the co-inventor of the bathysphere, in which he descended to a then record depth of 3,038 feet in the sea off Bermuda in 1934. Late one afternoon in a jungle glade in British Guiana, now Guyana, Beebe watched 63 individuals of two species of *Heliconius* butterflies, well known for their toxicity, settle down on their two separate nighttime roosts: "one by one they alighted on the very tips of bare twigs, upside down with closed wings . . . When I disturbed them, they flew up in a colorful flurry, flapped about for a minute or less, and returned, each to its particular perch . . . This persistent choice of position was invariably the case, as I observed in a number of butterflies which had recognizable tears in their wings. No matter how often they were disturbed, they never made a mistake." They used the same roosts for several weeks.

There are two possible explanations for how these butterflies and other insects relocate their roosts: As often is the case, especially with short-lived species, the insects may mark their nighttime roosts with a pheromone that can be smelled at a distance. Robert and Janice Matthews pointed out that pheromones that promote aggregations are particularly prevalent among aposematic insects—such as monarchs, probably ladybird beetles, and, as Thomas Eisner and Fotis Kafatos reported, the noxious yellow and black beetles known as lycids. The second explanation, put forward by John Turner, is that insects, such as *Heliconius* butterflies, that survive for weeks or months live so long that they could learn by sight the location of their communal sleeping sites and thus have no need of a marker pheromone. Working in Florida with the only *Heliconius*, the zebra, that regularly occurs north of Mexico, Frank Jones demonstrated that these butterflies can, indeed, relocate their communal roost without the aid of a marker pheromone. He cut away and removed all of the branches of a sleeping site and "brought others from a distant brush-heap, tying them into place as nearly as possible in the old site. At dusk, without any obvious hesitation, the butterflies accepted this substitution, and that night they slept upon the new [and certainly pheromone-free] twigs in their usual numbers."

❁ ❁ ❁   Truly aposematic animals often mimic each other in appearance and behavior, a phenomenon known as Müllerian mimicry in honor of its discoverer, Fritz Müller. Müllerian mimics are not bluffing; they are all noxious to predators in one way or another. In an 1879 article, Müller argued that different noxious and warningly colored species will often resemble each other because they can benefit by evolving similar warning signals, by adopting similar advertising logos. Some individuals of even the most toxic species will be killed in the process of educating and reeducating predators. If two or more species have the same or similar warning signals, the members of all of these species will benefit. There is only one signal for predators to learn, and since they cannot distinguish between the different species in a Müllerian complex, the inevitable mortality will be shared among a larger number of individuals.

For example, most of the insects that feed on milkweed, including monarchs and various species of aphids, beetles, and bugs, share red and black color patterns. They are Müllerian rather than Batesian mimics of each other, since they—at least all that have been checked so far—are noxious because of the toxins that they obtain from the milkweeds they eat. (Monarch caterpillars are certainly noxious and conspicuous, but are differently marked with black, yellow, and white rings around the body.) Different species of the toxic and warningly colored *Heliconius* butterflies of the New World tropics so closely resemble each other that they can be told apart only in the hand. Most of the thousands of species of wasps, protected by the sting of the females, constitute a huge complex of Müllerian mimics that are marked with yellow and black.

Woodruff Benson took advantage of the communal night roosting of a Costa Rican species of *Heliconius* butterfly to demonstrate the value of Müllerian mimicry by altering their color pattern. Working at night, he changed the appearance of some members of each of several roosts by

obscuring a red band on each forewing by staining it with black to match the rest of the wing, thus obliterating these individuals' resemblance to the other members of their species. The color of control butterflies was not changed but an equal amount of black stain was applied to a black area of each forewing as a control for the effects of handling and any possible toxicity of the stain. Although only one species of butterfly was involved, Benson's experiment is presumably a fair test of Müllerian mimicry, because the members of a species all have the same pattern and are thus perfect "Müllerian mimics" of each other. The results were striking and left no doubt that altering the pattern of one of these butterflies decreased the probability that it would survive. By checking their roosts at night, Benson found that butterflies whose color had been altered did not survive as long as the controls and sustained more wing damage, apparently inflicted by birds, than did control butterflies.

⊛ ⊛ ⊛   While some insects, such as the ladybird beetles that I have already discussed and the monarch butterflies that I will discuss in detail in the next chapter, form huge overwintering aggregations, other insects overwinter singly or in smaller groups. Among those that form small aggregations are the overwintering queens of certain social wasps. Unlike colonies of honey bees, ants, and termites, temperate zone colonies of social wasps—and bumble bees, too—are deciduous: they die off in autumn and new colonies are founded in spring by mated queens that survive the winter by diapausing in sheltered nooks and crannies. Bumble bee queens and some wasp queens overwinter as lone individuals, but among the queens of social wasps that overwinter in groups, are those of some species of the genus *Polistes*, whose uncovered paper combs are often seen hanging from the ceilings of outbuildings or the eaves of houses. In 1941, Phil Rau reported that swarms of these wasps "may be seen in various places away from the nests during the chilly nights and cloudy days of autumn, before they finally enter permanent winter quarters." He also found closely packed clusters of *Polistes* wasps in the dead of winter in hollow logs and other shelters. Sometimes they get into spaces in the walls of buildings and alarm people when they blunder into occupied areas of the building, as once hap-

pened in the Psychology Building at the University of Illinois in Urbana-Champaign.

Polistes are certainly well protected against predators. I know from
personal experience that their stings are extremely painful. Once while
mowing my lawn I was stung on my bare arm by a *Polistes* female from
a nest that, unknown to me, hung in a sheltered spot under the frame of
a basement window. I hadn't seen the wasp, and the pain was so sudden and fierce that I thought I might have been burned by a blob of mol-
ten metal thrown off by my power mower. After determining that nothing was amiss with the mower, I continued cutting the grass and was
again stung by a wasp when I passed near that same window. This time
I saw the wasp and discovered the nest.

Studies of stings, well summarized by Justin Schmidt, show that the
stings of social wasps are usually more painful to humans, and presumably other vertebrates, than are the stings of solitary wasps. This is to be
expected since the plentiful brood in the nest of a social species is a large
food resource that is likely to attract vertebrate predators ranging in size
from mice to bears.

❀ ❀ ❀   In July of 1978 I encountered a small but ominous-looking
school of tiny saltwater catfish while wading in waist-deep water in a
quiet bay near Batangas on the island of Luzon in the Philippines. The
fish were almost entirely black, and there were lots of them, several
hundred, swimming in such tight formation that the little school looked
like a black, basketball-sized amoeba undulating through the clear water. The school was very conspicuous and made me think of William
Henry Hudson's description of the feeding bands of closely packed
black grasshoppers he saw in Argentina. I knew about aposematic signals, and I knew that catfish have sharp spines, which are sometimes
venomous, in their dorsal and pectoral fins. It was obvious that these
tiny fish were sending me a warning. Consequently, I did not touch
them, although I stayed close by to watch them.

Several years later I found an article by Thomas Mortensen, who,
about 60 years before me, had encountered the same little fish, which he
identified as *Plotosus anguillaris,* on a coral reef near Zamboanga, which
is also in the Philippines, but on the island of Mindanao about 500 miles

south of Batangas. Mortensen did not heed the fishes' warning—a big mistake, but a convincing demonstration that aposematic animals profit by being gregarious. He scooped up most of the school with a single stroke of a hand net and proceeded to take them from the net with his bare hand. He wrote: "The first one I touched stuck to my fingers, producing a most intense pain, and on trying to get it off, I had it hanging in my other fingers. It was exceedingly painful, the pain lasting quite a while after I had succeeded in getting it off. After this experience I avoided, of course, most carefully to touch any specimen of this fish, and . . . I kept carefully away from these black, rolling balls."

⚙ ⚙ ⚙   But the ultimate broadcasters of group warnings are monarch butterflies, the subject of the following chapter. They require a chapter of their own because so much is known about the physiological, behavioral, and ecological aspects of their chemical defense. Monarchs are unique among the insects of the world in that they spend the winter in huge and dense aggregations that often include many millions of individuals that crowd onto just a few acres of land and virtually cover the trees growing there. Monarchs are poisonous and, like the insects and other creatures you met in this chapter, advertise their toxicity—in their case with a color pattern of bright orange and black and the bitter flavor of the scales that cover their bodies. But some birds have evolved ways to eat the monarchs in these overwintering aggregations despite their toxicity, and in the next chapter we will discover how group living allows monarchs to survive the depredations of these birds and how it confers other advantages on these spectacular butterflies.

Millions of Monarchs

A cluster of migrating monarchs resting overnight
on the branch of a maple

✦ ✦ ✦   On a winter day in 1974 Ken and Cathy Brugger came upon dead and tattered monarch butterflies along the side of a road as they drove through the high mountains to the west of Mexico City. This was an exciting discovery, a clue to the whereabouts of the long-sought-for winter home of these lovely orange and black migrants from farther north. An exploration of the surrounding countryside finally revealed an almost unbelievably dense aggregation of monarchs, countless millions of them literally blanketing a small grove of cypresses, firs, and pines. So tightly packed were the butterflies that the trunks, branches, and foliage of the trees were almost totally obscured by clusters of butterflies perched so closely together that their bodies touched. When it was cold and cloudy, they hung from the trees with their wings held tightly together up over their backs, only their pale undersides showing. The trees then seemed to be covered with pale dead leaves. When the sun shone, the butterflies spread their wings to bask in its warmth, revealing their orange and black upper sides and painting the trees bright orange.

As Frederick Urquhart wrote in a 1976 article in *National Geographic,* this discovery was the culmination of a 38-year search for the winter home of the monarchs of eastern North America. These butterflies are a beautiful and welcome presence during the summer in southern Canada and throughout much of the United States, but by winter they have all disappeared from their breeding areas—not a single egg, caterpillar, pupa, or adult butterfly is to be found. In the nineteenth century many scientists believed that monarchs overwinter within their northern breeding range, hidden in hollow trees and similar sheltered places. But by the beginning of the twentieth century opinion had begun to shift. Despite much searching, no diapausing monarch eggs, caterpillars, pupae, or adults were ever found in the north. It seemed likely that these butterflies migrate to warm areas to the south of their breeding range.

People have long known that monarchs from west of the Rocky Mountains spend the winter in aggregations of thousands of individuals clustered in Monterey pines and eucalyptus trees at 40 or more sites along the California coast from Monterey south to Los Angeles. But until the Brugger's discovery, no one knew what happened in winter to the monarchs that breed east of the Rockies. It was generally assumed that they migrated south, but no large overwintering populations of eastern monarchs had ever been found. A few monarchs are present in winter along the coast of Florida and on islands in the Gulf of Mexico, but these populations are wiped out in severe winters. Like many other entomologists, Urquhart, a professor of zoology at the University of Toronto in Canada, wondered where the monarchs went. He was determined to be the one to find out. In 1937, he and his wife, Norah, decided that the most reasonable approach to this problem was to track the southbound monarchs by releasing tagged individuals that they hoped would be found and returned by people along the migration route. But how does one tag a featherweight butterfly? They experimented with light-weight paper labels and several different kinds of adhesive, finally settling on a label that was glued to a part of the wing from which the scales had been removed and that bore an identifying code and the words "Send to Zoology University Toronto Canada." As it turned out, they would devote a lifetime to this project. Over the years, thousands of collaborators from all over North America joined the Urquharts in tagging monarchs. Hundreds of thousands of them were tagged and released, and as Urquhart wrote, tagged specimens were returned from "Maine and Ontario to California and Mexico, from Florida to the shores of Lake Superior."

As the accumulating returns were mapped, it gradually became apparent that the great majority of eastern monarchs move in a southwesterly direction to funnel through Texas on their way to Mexico. (A few may fly south along the Florida peninsula and cross the sea to the Yucatan by way of Cuba, but this route remains hypothetical.) The trail of marked monarchs ended in the high mountains west of Mexico City with the recovery of a few tagged individuals, but no overwintering site was found until after Ken Brugger, an American resident of Mexico City, responded to a request for volunteer help that Norah Urquhart had sent to Mexican newspapers in 1972. Brugger searched for monarchs as he drove his motor home along mountain roads in the area where tagged individuals had been found. In 1973 he saw monarchs being pelted out of the sky by hail, indicating that he was probably searching in the right area. He was later joined by his new bride, Cathy, and it was not long before they together found the overwintering site that I have described.

Although the Urquharts published accounts of the site, both in the *National Geographic* and the *Journal of the Lepidopterists' Society,* they did not reveal its location, and made the misleading statement that it is located in the Sierra Madre Mountains. The Urquharts refused to divulge the location of the site, not even to Lincoln Brower, then a professor at Amherst College in Massachusetts. At that time he was one of the leading authorities on monarchs and is now recognized as *the* leading authority on these butterflies. Using topographic maps and two clues gleaned from Urquhart's publications, Brower and his cooperators soon located this site and by 1986 a total of 12 such sites. In 1995, Brower published a lengthy and comprehensive account of the history and the current status of our knowledge of the monarch's migration.

❀ ❀ ❀ In 1977, Brower and several of his colleagues published their first scientific account of this site, which they had designated site alpha. They found that the monarchs, about 14.25 million of them, occupied an area of only about 3.7 acres. This small area included about 2,375 trees, about 97 percent of them firs, pines, and cypresses. The fir, known as the *oyamel* in Mexico, is by far the dominant member of this forest association, which is technically known as the *oyamel* fir forest ecosystem. This ecosystem consists of only 13 islands of vegetation high on some of the taller mountain peaks of Mexico; 9 of these islands are in the Transverse Neovolcanic Belt near Mexico City. All of the overwintering sites known as of 1995 are in *oyamel* forests at elevations of from about 9,800 feet to about 11,000 feet above sea level on peaks in nine different mountain massifs that are from about 43 to 106 miles west of Mexico City in the states of Mexico and Michoacan.

Although the sites are within the tropical zone, their climate resembles that of coniferous forests much father north in the United States and Canada. At that high elevation, the weather is cold in winter; frosts are not uncommon; and there are occasional snowfalls. On a day in January, Brower and his colleagues found the dawn temperature to be at

the freezing point in a clearing near the site, but within the *oyamel* forest where the butterflies clustered, the minimum temperature on that day was about 43°F. During a period of several days in the same month, the temperature sometimes fell below freezing in clearings, but within the site itself, it ranged from a low of about 42°F to a high of about 60°F. The forest obviously has an ameliorating effect that protects the monarchs from freezing. They concentrate in the most favorable part of the forest and shelter from stormy winds by not resting in the tops of the tallest trees, which rise above the surrounding trees and are thus exposed to gale-force winds. Most of the monarchs cluster 6 feet or more above ground level, thereby avoiding fallen snow and the lowest layers of cold air that might on occasion dip below the freezing point.

The majority of the monarchs are to be found resting on the *oyamel* firs, which are by far the most numerous trees and thus offer the most space to the butterflies, but the densest concentrations are on the less numerous cypresses, which seem to offer the best footing. As Brower put it, "the cypresses are often bowed over under the weight of the monarchs."

Most of the time it is too cold for the overwintering monarchs to become active but not cold enough to freeze them to death. It is warm enough so that they can maintain their grip on the trees and thus keep their clusters intact, but not so warm as to cause excessive activity. As Urquhart and later Alan Masters, Stephen Malcolm, and Brower reasoned, this nearly continuous state of semi-dormancy makes the survival of these butterflies possible, because their demands for energy remain low, and they thus conserve the reserve fat they will later need to fuel the first lap of their return flight to the north. It is wet enough in the forest to keep the rate at which the monarchs lose precious water from their bodies at a low level but not so wet as to preclude all activity. On warm days the overwintering monarchs fly to creeks or soggy soil to drink and replenish the water content of their bodies, sometimes traveling for more than a half mile.

A mere handful of lucky individuals may appropriate the nectar from the few flowering plants that grow nearby, but there isn't nearly enough to feed millions of butterflies, and almost all the members of a colony must do without food for the entire winter. Lincoln Brower told me that when he and the botanist Chris O'Neil examined monarchs at one of the

Mexican sites, they found that individuals that stayed on the trees were mainly fat and healthy but had empty crops. They had obviously not obtained any nectar from the nearby plants, but were still fat from the ample nectar that they had ingested before arriving at the *oyamel* forest. The butterflies that had flown to flowers near the overwintering site, mainly thin and apparently unhealthy individuals, had full crops. But their crops were filled with air, not nectar! In a vain quest for food, these starving butterflies had been sucking air from blossoms that had already been emptied of their nectar.

North American monarchs of the generation that matures in late August and migrates to Mexico are, unlike those of earlier generations, in a *reproductive diapause.* Loosely speaking, you could say that they are in a "partial diapause," which is induced by the relatively short days and lower temperatures of late August. Although these butterflies in reproductive diapause do not reproduce—they seldom mate and if they do their eggs do not develop—they do remain active, capable of drinking nectar and making the long and energy-consuming flight to Mexico. Along the way they often cluster on trees or bushes at night, forming temporary sleeping aggregations that may include from a dozen or so to hundreds of individuals. They stop to take nectar throughout their southward trip, but in Texas and northern Mexico they really load up from fall-flowering plants, and by the time they arrive at their winter sites in mid-November, their bodies consist of about 50 percent fat, enough to supply the energy they will need during the winter. Recently, David Gibo and Jody McCurdy did experiments that indicate that migrants who leave Ontario for the south at the beginning or middle of the 2-month migration period contain large quantities of body fat and are likely to reach Mexico. Those that leave toward the end of the migration period contain only small quantities of fat and are far less likely to make it to Mexico.

Until recently, no one knew how monarchs find their way to their distant overwintering sites. But research done in the past few years by Sandra Perez, Orly Taylor, and Rudolf Jander of the University of Kansas indicates that these butterflies, like many migrating birds, navigate by means of both a "sun compass" and a "geomagnetic compass."

Since the sun passes across the sky from dawn to dusk, its position tells us the direction only if we know what time it is. Many animals, in-

cluding insects, can tell time by means of an internal biological clock that is physiologically driven. A biological clock can be reset by keeping an animal under artificial illumination that goes on and off to create artificial "dawns" and "dusks" that come earlier or later than natural dawn and dusk.

Perez and her fellow researchers reset the biological clocks of a large number of migratory monarchs so that they ran 6 hours slow. When these butterflies were released, they—duped by the false information from their biological clocks—flew toward the west rather than toward the south-southwest as did wild monarchs and experimental controls subjected to artificial "dawns" and "dusks" that came at the same time as natural dawn and dusk. The obvious conclusion is that monarchs can tell direction from the position of the sun.

But even on overcast days when the sun is not visible, migratory monarchs manage to fly in the correct direction. As Perez and her co-workers wrote, this indicates that these butterflies "have a non-celestial backup mechanism of orientation, such as a geomagnetic compass." In very recent experiments, these researchers demonstrated that monarchs do have a geomagnetic sense that could be used as a compass. Part of a large group of southward-migrating monarchs that had been captured and held in an outdoor screenhouse for several days were exposed to a strong magnetic field just before being released. They were disoriented and flew off in random directions. But the other part of this group, not exposed to a magnetic field, were not disoriented when released and flew off in the correct direction, toward the south-southwest.

❀ ❀ ❀   When the time to leave the winter site and return to their breeding range in the north approaches, the monarchs' reproductive diapause is terminated by a combination of the warmer air temperatures and longer days of late January and early February. As January progresses, maximum temperatures become increasingly warmer, sometimes getting up to the high 60s and low 70s Fahrenheit. By the first of February, the duration of the daylight period has increased to 11 hours and 18 minutes, only a half hour longer than the 10 hours and 48 minutes of December 21, the winter solstice and the shortest day of the year, but long enough to trigger the termination of diapause. In controlled laboratory experiments, John Barker and William Herman showed that monarchs remain in reproductive diapause at day lengths of 11 hours or less; their eggs do not develop, and, as is known from studies at overwintering sites, their desire to mate is inhibited.

As the days lengthen beyond the critical 11 hours, more and more mating pairs of monarchs are seen at the overwintering sites in Mexico. A few matings occur in January; the number of matings increases throughout the month of February; and by March mating has become intense. Coupled pairs are everywhere, and the butterflies swarm through the air, their bodies a dense blizzard of orange against the blue sky, in an unrestrained orgy as males pursue females. A description of mass matings at an overwintering site in California by H. Frederick Hill, Jr., and two colleagues probably applies equally well to monarchs in Mexico. As the time to disperse approached, the end of February in California, there were days on which nearly all of the butterflies were engaged in sexual pursuits. "On warm afternoons . . . the air was filled with thousands of rapidly flying butterflies twisting and turning in flight." Males did their best to force females to the ground. Typically, as Thomas Pliske noted in another article, the male lands on top of a flying female and clutches her body with his legs. "In so doing he wraps . . . his legs around and under her wings. [She] struggles to fly, but the male ceases flying although keeping his wings open in the sailing position so that the pair slowly fall to earth." The male then forces the female to copulate. Under less frenzied circumstances, a male sometimes takes a more seductive approach than forceful rape, flying just in front of the female as he extrudes his hairpencils, brushlike organs at the end of the abdomen, dipping and bobbing as he brings them in contact with her head and antennae. In species closely related to the monarch, the hairpencils shower the female's antennae with an aphrodisiac dust, but there is some doubt as to whether or not this happens in the monarch.

At the end of March or in early April, the monarchs leave the *oyamel* forests in the mountains of Mexico and move northward to repopulate their breeding range in the eastern United States and Canada. Along the way they drink nectar from blossoms and lay eggs on newly emerging milkweed shoots. Most of the males will die along the way, but until the end they chase females and copulate as often as possible. Since the name of the evolutionary game is the survival of the fittest, and the

fitness of an individual is measured by how many surviving descendants it leaves behind to pass its genes on to the future, the most useful thing an adult male monarch can do to increase his fitness is to inseminate as many females as he possibly can.

The female's fitness also depends upon the number of offspring that survive her. But mother monarchs, like most female animals, attain fitness by a different route than do the males and tend to be less promiscuous than males. First, a female's contribution of sex cells is far greater than a male's. A male contributes to his mate only sperm cells contained in a few tiny droplets of semen, a contribution necessary for reproduction, but a small one that requires little energy or other resources to produce. A female, by contrast, contributes to each of her many offspring a yolk-filled egg that is immensely huge in comparison to a sperm cell, and whose production requires far more energy and other resources.

Second, the mother monarch gives her eggs a modicum of what we can, perhaps loosely speaking, call parental care. When we think of parental care, we picture a worker honey bee feeding a helpless larva, a bird incubating its eggs, or a human mother lifting her baby from its cradle and putting it to her breast. But there are simpler and less comprehensive forms of parental care, such as the careful and appropriate placement of their eggs practiced by many female insects. They don't lay their eggs just anywhere, although they immediately abandon them. Bluebottle flies lay them on the dead bodies that their larvae will consume. Parasitic wasps inject their eggs into the bodies of the insects that will be the hosts of their larvae. Plant-feeding insects, including monarchs, lay their eggs on species of plants that their larval offspring will be willing to eat. The importance of this last form of care becomes apparent once we realize that many plant-feeding insects are highly host-specific: they will eat only a few closely related plants and will starve to death if offered nothing but other plants that are not on their menu. Just as tomato hornworm caterpillars will eat only the leaves of tomato, tobacco, or other members of the nightshade family, monarch caterpillars will, with only a few exceptions, eat only the foliage of various species of milkweed, mainly members of the genus *Asclepias* of the family Asclepiadaceae.

As far as I know, no one has determined the average number of eggs that female monarchs lay during their lifetime. But as Fred Urquhart reported in his book on the monarch, more than 400 developing eggs have been counted in the ovaries of one female. A female might possibly lay that many eggs, but in reality the total actually laid is probably considerably less. It does not, however, stretch the imagination to think that monarchs may lay a minimum of 100 or 200 eggs.

The females spend some time seeking the sugar-rich nectar that fuels their flight and mating with males, sometimes as many as seven of them but usually fewer. But they spend most of their time cruising through the air as they search for milkweed plants. The eggs are laid one at a time, glued to the underside of a leaf. Only one egg is laid on a leaf, and only a few eggs are laid on any one plant. In the warm weather of midsummer, the embryo in the egg develops rapidly, and the caterpillar (larva) chews its way through the egg shell after only 3 or 4 days. During the cooler weather of early and late summer, embryonic development may require as much as 6 days.

While the adult butterfly has coiled soda-straw–like mouthparts ideally suited for sipping nectar, the larva's mouthparts are designed for snipping and chewing solids. The newly hatched larva first eats the shell of its egg, but after that eats nothing but the leaves—or occasionally the blossoms—of the milkweed plant on which it was placed as an egg by its mother. The caterpillar takes about 2 weeks to complete its growth, molting its skin four times to accommodate a tremendous increase in size as it multiplies its weight by a factor of between 2,000 and 3,000. The full-grown caterpillar usually leaves the plant in search of a suitable pupation site when it is ready to molt to the pupal stage, the developmental stage in which the caterpillar metamorphoses into a butterfly, known as the chrysalis to butterfly collectors. It will settle on almost any favorable site, such as the underside of a tree limb, the eave of a house, or the lower surface of a leaf of almost any kind of plant. It spins a thick mat of silk and hangs from it upside down by hooks on the legs at the end of its abdomen as it molts its larval skin to reveal the pupa. Before it completely separates from its larval skin, the pupa embeds in the mat of silk the spines on a club-shaped protrusion from the end of its abdomen. The beautiful pupa, blue-green and studded with golden spots, then hangs suspended upside down for about 2 weeks as the marvelous transformation to a butterfly takes place.

After that, the adult emerges from the pupal skin, hangs upside down

for a time as its wings expand and dry, and finally flies off. As you know, adults that emerge in late August are in reproductive diapause, will migrate to Mexico, and will not lay eggs until they move north from Mexico the following spring. Monarchs that emerge earlier in the summer or in the spring do not enter diapause and commence reproductive activity within days of their emergence. There will be several generations during spring and summer, the number varying with the latitude.

❀ ❀ ❀   By late spring, the monarchs have repopulated all of their breeding range from the southern United States to Canada. There are two hypotheses, summarized by Brower in 1995, as to how this is accomplished: The "single-sweep hypothesis" is that the northward migrants pause to lay eggs in the southern United States but then fly directly to the more northern parts of their breeding range, including Canada. The opposing "successive-brood hypothesis" is that the returnees from Mexico fly to the Gulf Coast, lay eggs there, and then die. It is their progeny that continue the migration northward to southern Canada.

Brower's research group found strong support for the successive-brood hypothesis by "fingerprinting" monarchs at different times and places during their yearly cycle. The fingerprint, more specifically the cardiac glycoside fingerprint, is a chemically analyzed measure of the different kinds and quantities of cardiac glycosides, also known as cardenolides, that are present in a monarch's body. As you will read below, cardenolides, so called because they are used to treat heart problems in humans, are the toxins that monarch caterpillars obtain from the milkweeds they eat and store in their bodies through to the adult stage. Since different species of milkweed plants contain different kinds and quantities of these substances, and since these differences are reflected by the cardenolide content of the adult butterfly, a monarch's cardiac glycoside fingerprint reveals which species of milkweed it ate when it was a caterpillar.

The fingerprints showed that, of several hundred monarchs collected as they migrated southward, rested at their overwintering sites in Mexico, or first arrived at the Gulf Coast on their return journey from Mexico, over 80 percent had fed on common milkweed (*Asclepias syriaca*) as caterpillars. These butterflies, all from the overwintering population, had clearly originated in the northern United States and southern Canada, since common milkweed is abundant there but absent or rare south of North Carolina and Kansas. In contrast, only 6 percent of over 600 monarchs collected in the northern United States in June—they could only have been returnees from the south—had fed on common milkweed. The rest of them, 94 percent, had cardiac glycoside fingerprints indicating that they had spent their caterpillar stage on one of several species of milkweed that are common in the southern United States but are rare or absent in the north. Clearly, the northern part of this butterfly's breeding range is repopulated each year not by individuals who arrive directly from Mexico but mostly by their progeny who grew up along the Gulf Coast. But on the basis of cardiac glycoside fingerprints and other evidence, Brower believes that a few monarchs may manage to make it all the way back to Canada from Mexico.

❁ ❁ ❁ The migration and the winter gatherings of the monarchs are among the most spectacular and awesome of all natural phenomena, unique in the insect world. Lincoln Brower wrote of his feelings on a warm March morning as he watched tens of thousands of these butterflies explode from their resting places on the trees at an overwintering site in Mexico: "Flying against the azure sky and past the green boughs of the oyamels, this myriad of dancing embers reinforced my earlier conclusion that this spectacle is a treasure comparable to the finest works of art that our world culture has produced over the past 4000 years." But even as I write this paragraph, the winter gathering places of the monarchs are being destroyed by illegal logging—indeed, all of the *oyamel* forests in Mexico are threatened by legal and illegal logging. If the logging continues at its present rate, all of the overwintering sites in Mexico will be gone by the first decades of the twenty-first century. So desperate is the situation that the Union for the Conservation of Nature and Natural Resources has recognized the monarch migration as an endangered biological phenomenon and has designated it the first priority in their effort to conserve the butterflies of the world.

All efforts to preserve the overwintering sites in Mexico have failed. In August of 1986, the Mexican government issued a proclamation des-

ignating these sites as ecological preserves. Five of the 12 known sites were to receive complete protection. Logging and agricultural development were to be prohibited in their core areas, a total of only 17 square miles, and only limited logging was to be permitted in buffer zones surrounding the cores, a total of another 43 square miles. The proclamation is largely ignored. One of the 5 "protected" sites has been clear-cut, some buffer zones have been more or less completely destroyed, and trees are being cut in all of the core areas. As Brower told me, guards that were appointed to protect the monarch colonies have not prevented illegal logging but have barred tourists, film crews, and scientists from witnessing logging activities. It is incomprehensible to me that a way cannot be found to protect a mere 60 square miles of land that are home to one of the world's most spectacular biological phenomena.

If the monarchs are to survive, the *oyamel* forest in which they spend the winter must remain intact. Even minor thinning of the core areas causes high mortality among the butterflies, because the canopy of the intact forest serves as a protective blanket and umbrella for them. Within a dense stand of trees, the temperature does not drop as low as it does elsewhere, enabling the monarchs to survive freezing weather under the blanket of trees. Thinning the trees puts holes in the "umbrella" that protects the monarchs, letting them get wet during winter storms. A wet butterfly loses its resistance to freezing and dies. Even a dry butterfly loses precious calories as its body heat radiates out to the cold night sky through holes in the canopy.

But how secure is the monarch population on its breeding range in the United States and Canada? Probably not as secure as we might imagine. Leonard Wassenaar and Keith Hobson used naturally occurring chemical markers—different forms of the hydrogen and carbon molecules that monarchs ingest with their food plants—to determine where overwintering monarchs had come from, where they had fed and grown as caterpillars. In 1997, and probably in other years as well, about half of them had come from only a small part of the breeding range, a band about 400 miles wide that extends from Kansas and Nebraska east to Ohio. This area includes most of the corn belt, and is intensively treated with herbicides to kill a variety of "weeds," including the milkweed that is the primary food plant of monarchs. Improved methods of weed control could eradicate much of the milkweed in this area and thus take a heavy toll on the monarch population that migrates to Mexico.

There is a new threat to monarchs in the corn belt. John Losey and two coworkers recently discovered that pollen from genetically modified corn kills monarch caterpillars. A bacterial gene inserted in corn plants causes them to produce Bt, an insecticide naturally synthesized by the bacterium in question. The plants are thus toxic to insects that feed on them, including pestiferous European corn borer caterpillars. This serves the farmer but is a threat to the environment. After the corn pollen dries out, it is blown away by the wind, and has been found on leaves 200 feet from the closest corn plant. Corn produces pollen at the same time monarch butterflies are feeding. The pollen lands on nearby milkweed leaves and is swallowed by the caterpillars as they eat the leaves. In an experiment done by Losey and his colleagues, about half the caterpillars that had been fed leaves with Bt pollen died after 4 days, but all that had been fed leaves with Bt-free polllen survived. The threat to monarchs is serious, because the acreage planted with Bt corn is rapidly increasing, and milkweeds have few refuges more than 200 feet from corn plants in the intensively cultivated corn belt. Wind-blown Bt pollen surely lands on the leaves of plants other than milkweeds, and thus is also a threat to many other species of insects, especially caterpillars, that eat these leaves.

❀ ❀ ❀   It is now a well-established fact that the bodies of many monarch butterflies, but not all, are noxious, laced with toxic substances that make insect-eating predators such as birds ill soon after they ingest a toxin-containing individual. These toxic substances in monarchs are the cardiac glycosides, or cardenolides, that I have already mentioned. As you know, many other creatures, but mainly insects such as some other butterflies and ladybird beetles, are noxious to predators because of toxins in their bodies. Some synthesize their own toxins, but others, like the monarch, sequester toxins that they obtain from their food plants. Plants manufacture toxins to prevent insects and other herbivores from feeding on them. The monarch, like some other insects, has turned the table on its host plant. Not only has it evolved a way to cope with the milkweeds' cardenolide toxins, but it has also evolved a way to

use them against its own enemies. Monarchs alert predators to their noxiousness by means of their strikingly conspicuous orange and black coloration and, if actually seized by a predator, by the bitter flavor of the cardenolides that they contain, which can be tasted on contact and are most concentrated in the wings, the "handles" by which birds first grab butterflies.

There have been several experimental demonstrations of the effectiveness of warning signals, including the one with lizards and toxic butterflies that I discussed in the previous chapter. But my favorite, and the most revealing of them all, is one done by Lincoln Brower in 1969 with blue jays and monarch butterflies. The cardenolides that many monarchs contain are highly toxic and can be lethal, but fortunately for blue jays and probably other creatures that may try to eat monarch butterflies, the toxic dose is somewhat higher than the emetic dose. Thus a predator that eats a monarch will probably vomit and purge itself of the cardenolides before they can kill it.

In his experiment, Brower first captured wild blue jays and kept them in cages until they were willing to sample monarch butterflies, presumably having forgotten whatever they had learned in the wild about the noxiousness of these insects. Then he offered to these "brainwashed" blue jays monarchs that did not contain cardenolides because they had been raised in captivity on a species of milkweed that does not contain these toxins. The blue jays readily ate these cardenolide-free butterflies and happily continued to eat them until after the next step in the experiment. The next step, as you have probably figured out, was to offer these same blue jays monarchs that were toxic because they had been raised on a species of milkweed that does contain cardenolides. The jays ate these toxic monarchs but almost immediately showed signs of distress, raising their crests and fluffing out their feathers. Soon thereafter they vomited, as many as nine times during a half-hour period. This turned out to be one-shot learning. Blue jays that had suffered from eating a toxic monarch thereafter refused to so much as touch either toxic or toxin-free monarchs. Some of them even retched at the mere sight of a monarch.

Certain observations done in nature strongly support Brower's findings by showing that birds seldom attack wild monarchs not reared in captivity. It would be next to impossible to catch a free-ranging bird in the act of spotting a monarch and then refusing to attack or eat it. The next best approach is to examine the wings of wild monarchs for signs of damage inflicted by the beaks of birds. Brower and his colleagues caught and examined many wild monarchs. Most of them showed no trace of injury, and the few that did showed only the crisp and unmistakable print of a beak, indicating that they had been grabbed by a bird and immediately released, presumably in response to the bitter taste of the cardenolides on the wing.

James Sternburg and I made similar observations in central Illinois. We caught and examined several hundred monarchs during their southward migration through Illinois. Like the Brower group, we found that only a very few monarchs showed any sign of having been attacked by a bird, or any other sort of injury, and that was almost always nothing more than the imprint of a bird's beak. But we discovered something new. Most of the butterflies that bore a beak mark, or sometimes more than one beak mark, were abnormal individuals with one more or less shrunken hind wing. Almost all of these abnormal individuals bore at least one beak mark. They managed to fly, but they probably flew in an abnormal manner, with a "limp" that attracted the attention of insect-eating birds.

In 1976, Eberhard Curio, a well-known German animal behaviorist, reported several observations of predators preferentially attacking injured prey, presumably because they perceive them as being different and perhaps easy to catch and kill. It may be that in the minds of the birds that attacked monarchs, the prospect of an easy kill outweighed the warning of noxiousness. In a booklet he published in 1957, Fred Urquhart noted that although untagged monarchs were seldom if ever attacked by birds, tagged monarchs were often attacked. It seems likely to me that the tagged monarchs were picked on because they were perceived as being different and perhaps good to eat.

❁ ❁ ❁   Not all monarch individuals are protected by noxiousness. Broadly speaking, there are two reasons for this. First, because of differences in the quantity and quality of cardenolides in the various species of milkweeds that the caterpillars eat, the cardenolide content of adult monarchs varies to such an extent that some are much less noxious than

others and some are not noxious at all. Second, some birds and mice have evolved ways to overcome the monarch's chemical defense. At least one bird, the black-headed grosbeak, can eat these butterflies with relative impunity, because it is physiologically resistant to cardenolides. A few others are able to detect and reject noxious monarchs and eat only the relatively innocuous ones or to selectively ingest the least noxious parts of their bodies while rejecting the more noxious parts.

There are 108 species of milkweed in North America, and at least 28 of them are eaten by monarch caterpillars. These 28 species, however, differ greatly from each other in the cardenolide content of their leaves. The difference between the sandhill, or purple, milkweed (*Asclepias humistrata*) and the butterfly weed (*Asclepias tuberosa*) makes this point nicely. Both species occur throughout much of the southeastern United States, often growing together at the same site, and both are regularly eaten by monarch caterpillars. The leaves of the sandhill milkweed are rich in cardenolides, and one adult monarch that was raised on this plant contained enough of them to make 14 blue jays vomit. In contrast, there are virtually no cardenolides in the leaves of the butterfly weed, and adult monarchs that fed on it as caterpillars are readily eaten by naive blue jays and do not cause them to vomit.

In a fascinating and beautifully illustrated article in *Terra,* the magazine of the Natural History Museum of Los Angeles County, Lincoln Brower related how he collected the monarchs that Jane Van Zandt Brower used in the first experiments that demonstrated that monarchs are noxious to birds, Florida scrub jays in this case. By good luck, he had collected the monarchs at a site where they had been feeding on the cardenolide-rich sandhill milkweed. If he had collected them from some other site where only the butterfly weed grew, they would have been palatable to the jays, and Jane Brower's findings would have supported Fred Urquhart's ill-founded contention that monarchs are never noxious, supported only by anecdotal accounts that birds sometimes eat monarchs and by the fact that when he and some of his students ate these butterflies they did not find them to be bitter.

The common milkweed (*Asclepias syriaca*) grows throughout most of the northeastern quarter of the continental United States and in adjacent southern Canada. Almost everywhere throughout its range, it is one of the most abundant species of milkweed, and in many areas disturbed by agriculture it is by far the most common species present, as is especially obvious in late summer, when it releases its plumed seeds. As Stephen Malcolm of Western Michigan University has noted, common milkweed varies greatly in its cardenolide content from plant to plant and from place to place. Consequently, monarch butterflies that fed on this species as caterpillars vary in their noxiousness to birds—from individuals that are palatable and readily eaten to those that cause birds to vomit.

The upshot of all of this is that most of the monarchs that spend the winter in Mexico have a low cardenolide content—for the reasons that I have already stated and also because their cardenolide content decreases during their lengthy southward migration. Only about 10 percent of them contain enough of these toxins to cause a blue jay and probably most other birds of about the same weight who injest just one monarch to vomit. This might lead one to think that the overwintering monarchs, most of which have little or no chemical protection, will be readily eaten by birds, mice, or other predators and may well be on the verge of near annihilation.

But this is definitely not the case. Most of the insect-eating vertebrates that are known to occur near the Mexican overwintering sites do not eat monarchs. Of the 5 species of mice that are abundant near these sites, only 1 species feeds on monarchs. Of the 37 species of insectivorous birds that have been recorded from these sites, 25 have never been seen to attack monarchs; 8 species attack them only rarely; and only 4 species regularly attack them; but only 2 of these 4 species eat large numbers of them.

The overwintering monarchs in Mexico are clearly protected from most of the vertebrates that might eat them. There are two or more reasons for this. First, most of the overwintering monarchs do contain *some* cardenolides, on the average only about two and a half times less than the dose that will make a blue jay vomit. Thus although eating only one monarch containing a low dose of cardenolides might not cause a bird to vomit, eating several of them one right after the other would probably do so and thus teach the bird to avoid monarchs in the future. Second, a dose of cardenolides too low to cause vomiting may still subject a

bird to some less obvious form of physiological distress that might induce it to shun monarchs. This is known to be the case with blue jays. Third, a bird will probably be more impressed by just a few unhappy experiences with toxic monarchs than by many happy experiences with palatable monarchs, and will consequently refrain from trying to eat monarchs at all. In a laboratory experiment with starlings and palatable mealworm beetle larvae, Jane Van Zandt Brower found that even a small proportion of "noxious" larvae, rendered so by dipping them into a strong solution of quinine dihydrochloride, dissuaded starlings from eating any mealworm larvae at all and thus gave considerable protection to identical palatable "mimics," larvae that had been dipped only into distilled water. Palatable monarchs are, of course, perfect mimics of unpalatable monarchs.

The black-headed grosbeak, which also occurs in the western United States and Canada, and the black-backed oriole, once considered to be of the same species as the Bullock's oriole of the United States and Canada, are the only birds that regularly eat large numbers of monarchs at the Mexican overwintering sites. Flying together in mixed flocks of one or two dozen individuals, these two species of birds regularly make feeding forays into the overwintering sites. That they actually eat monarchs is apparent from the discarded wings and partially devoured butterflies that litter the ground, from observations of their feeding made by researchers using binoculars, and from the contents of the birds' stomachs. Both species discard the wings, as do most birds that eat moths or butterflies, be they palatable or toxic. All of the grosbeak stomachs examined contained identifiable monarch remains, whole abdomens and sometimes whole thoraxes. For reasons that will become apparent below, identifiable remains were difficult to find in oriole stomachs, but there was no doubt that they had eaten monarchs, because their stomachs did contain cardenolides and sometimes traces of monarch remains.

Each of these birds has its own distinctive way of circumventing the monarchs' chemical defense. The black-headed grosbeak has a physiological tolerance for cardenolides: it can ingest them because it is at least partially immune to their effects. The grosbeaks use their thick, heavy bills to dismember monarchs, snapping off the abdomens, which they swallow whole, and often also totally or partially consuming the thorax. The feeding behavior of black-backed orioles is very different. It has to be, because they are not at all immune to cardenolides. They manage to eat monarchs by rejecting those with a high toxin content and by ingesting only the least toxic parts of the ones they do eat. The orioles grasp the butterflies by the wings, which have a higher concentration of cardenolides than any other part of the body, and reject the most toxic ones on the basis of the potency of the bitter cardenolide flavor of the wings. A monarch that is acceptably mild in flavor is dismembered and eaten much the way a lobster is eaten by a person. The butterfly is held down by the bird's foot as the bird's long, thin, pointed bill delicately strips the body of its thoracic muscles and the contents of the abdomen, the body parts that are eaten because they have the lowest concentration of cardenolides. The exoskeleton, the outer body wall, relatively rich in cardenolides, is left intact and discarded, just as the hard shell of a lobster is discarded by humans.

In a 1985 article in *Evolution,* Lincoln Brower and William Calvert reported an estimate of the number of monarchs that birds ate during the 135-day overwintering season in a 5.6 acre site in Mexico. They made their estimate by counting discarded wings and partly eaten monarchs that were caught in over 80 nets, each somewhat more than 1 square yard in size, that had been suspended above the ground within and at the periphery of the overwintering site. The total number of butterflies killed at this site was calculated from the total area of the nets and the area of the whole site. The number of monarchs killed was large, an average of about 15,067 per day, or a total of about 2,034,000 during the season. But this number must be seen in perspective. This overwintering site harbored about 22,500,000 monarchs. Hence, the birds killed only about 9 percent of the butterflies that were present. Another study showed that mice kill only about a sixth as many butterflies as do birds, or about 1.5 percent of those present. Thus these predators together killed only about 10.5 percent of the overwintering monarchs, leaving almost 90 percent unharmed.

There are several reasons why more monarchs are not killed. One is that their noxiousness completely or almost completely protects them from many birds and other vertebrate predators that would almost cer-

tainly eat them if they were palatable. Unlike black-headed grosbeaks and black-backed orioles, these other insect eaters have not evolved a way of getting around the monarch's defense. Furthermore, there may even be a limit to how many of these butterflies grosbeaks and orioles can tolerate. If toxic levels of cardenolides build up in their bodies over time, they may have to switch periodically to foods other than monarchs. This possibility was suggested to Brower and Calvert by a cyclical variation in the number of butterflies eaten per day by these birds, determined by the number of discarded wings found on the site. There were periods of 4 or 5 days during which they ate many butterflies followed by periods of about the same duration during which they ate very few. Another reason why so many monarchs survive is that there are more than enough of them to saturate the appetites of the creatures that eat them. Thus the mice, the black-headed grosbeaks, the black-backed orioles, and the other minor predators in the area could kill or eat no more than a relatively small percentage of monarchs even if they ate nothing but these butterflies.

There is a tremendous difference in the responses of insect-eating birds and other vertebrates to monarchs in the overwintering colonies in Mexico and California. While the eastern monarchs that migrate to Mexico are frequently attacked, the western monarchs that migrate to coastal aggregation sites in California are seldom bothered by predators except for occasional attacks by chestnut-backed chickadees. Since the eastern and western monarchs contain about the same *quantity* of cardenolides, this difference in predation is presumably due to the fact that the two groups contain different kinds of cardenolides that differ in *quality,* in their toxicity to vertebrates. This comes as no surprise because eastern and western monarch caterpillars eat different species of milkweeds that differ in the array of cardenolides they contain. The emetic potency—the ability to cause vomiting—of the California monarchs is 4.3 times as great as that of the Mexican monarchs. It follows that, other things being equal, a bird eating California monarchs will vomit after eating far fewer of them than are required to make a bird eating Mexican monarchs vomit. Furthermore, a bird foraging in an aggregation in California is much more likely to encounter a monarch that will cause vomiting than is a bird foraging in an aggregation in Mexico. Only 10 percent of the monarchs in Mexico are potent enough to cause vomiting by themselves, as compared with 49 percent of those in California.

✸ ✸ ✸   The noxiousness and easily learned and recognized warning signals of monarchs presents an opportunity for other butterflies to warn away potential predators by mimicking the monarch. This opportunity has not been missed. The monarch is closely mimicked by the viceroy—so closely that the two are difficult to distinguish unless you get a close look. Viceroys, only distantly related to the monarch, look nothing like their own close relatives, whose wings are mostly black with a bluish or purplish sheen traversed by broad white "disruptive" bands that make the butterfly less noticeable by obscuring its shape. This is presumably similar to the ancestral appearance of the viceroy, what the viceroy looked like before it evolved to resemble the monarch. Of the many relatives of the viceroy in North and South America, only a few do not have this black and white pattern, and these few are all mimics of one or another species of noxious butterfly. The precision with which natural selection creates mimics is indicated by the fact that in southern Florida and the southwestern United States, where monarchs are rare, the viceroy resembles not the monarch but rather the much more abundant queen, a close relative of the monarch that also feeds on milkweed and is also noxious. Queens look quite a bit different from monarchs, and the subspecies of the queen that occurs in Florida looks different from the subspecies in the southwest. Florida viceroys look like Florida queens, and southwestern viceroys look like southwestern queens—testimony to how perspicacious birds are in identifying their prey.

The viceroy's close resemblance to the monarch is often cited as an example of Batesian mimicry; this view holds that the edible viceroy is bluffing, is imitating the warning signals of the noxious monarchs to ward off predators that would find it palatable. But the available evidence indicates that the viceroy may be a Müllerian mimic of the monarch, a mimic that benefits merely by joining forces with the noxious monarch and presenting similar warning signals. In the late 1950s, after presenting whole specimens to captive, wild-caught Florida scrub jays,

Jane Van Zandt Brower concluded from the responses of the jays that viceroys are more palatable than monarchs but less palatable than control butterflies that are not warningly colored and were readily eaten by the jays. She argued that the viceroy is neither a Batesian nor a Müllerian mimic but falls somewhere in between. More recently, David Ritland and Lincoln Brower reported on a new experiment with monarchs and viceroys, differing from Jane Brower's experiment in that the predators were captive, wild-caught, male red-winged blackbirds; that queens as well as monarchs were tested; and that the birds were presented only with the abdomens of the butterflies, and thus were compelled to accept or reject on the basis of taste without the complication of wing pattern cues that they might have learned before the experiment. The results suggest that viceroys are Müllerian mimics of monarchs. Monarchs and viceroys were about equally unpalatable to red-winged blackbirds and viceroys were even more unpalatable than queens. But it may be that other birds, including Florida scrub jays, find viceroys to be more palatable than monarchs.

❀ ❀ ❀   Both the adult monarchs discussed in this chapter and the adult ladybirds discussed earlier clearly benefit by spending the winter in large and closely packed aggregations—ladybirds by the thousands and monarchs by the millions. But aside from coming together, the individuals in these groups display little or no cooperation, and they certainly don't build themselves a shelter. In the next chapter, we will look at a rudimentary society in which the individuals cooperate in several endeavors, most notably the construction of a home for the whole group.

Teams of Tent Caterpillars

*Tent caterpillars crawling from the nest to feed
on leaves at the end of a branch*

⚙ ⚙ ⚙   As odd as it may seem, there is a five-way ecological relationship between telephone companies, eastern tent caterpillars, robins, wild black cherry saplings, and cuckoos. Fifty years ago, before it was overrun by the Eastern Megalopolis, Fairfield County, Connecticut—where I grew up—still had country roadsides that were left to grow wild and were lined with rows of telephone poles. The wires strung between the poles were a favorite perch for birds, including many robins. Justly famous as hunters of earthworms, robins also feed voraciously on small fruits.

Many plants entice birds to distribute their seeds by enclosing them in small, tasty fruits. The bird swallows the fruit whole and later, sometimes miles away, passes out the unharmed seed with its excrement. (The fruits of some plants contain laxatives to speed the fruit through the bird's digestive tract.) The fleshy fruits of cherry trees contain a single hard and indigestible seed, are small enough to be swallowed whole, and are conspicuously colored in red, yellow, purple, or black to attract the attention of birds, which have good color vision. When wild black cherries are ripe, robins and other fruit-eating birds descend upon them and have a feast. They digest the nutritious pulp of the fruit but pass the seeds in their droppings, sometimes as they perch on a telephone wire. The seeds germinate and the seedlings grow rapidly, probably aided by the fertilizing bird dropping. Soon there are numerous black cherry saplings, an unusually dense stand, growing under the wires along the roadside. They may persist for years—until the next time the telephone company clears away the "brush".

Black cherry leaves are the favorite food of eastern tent caterpillars, although they happily attack other plants of the same family (the Rosaceae), including apple trees. In some years, eastern tent caterpillars are exceptionally abundant, and virtually every cherry sapling along the roadside will contain one or more of the large, white, silken tents that colonies of these caterpillars spin in a major fork of the tree.

Insectivorous birds, mainly yellow-billed and black-billed cuckoos, gravitate to this abundance of insect food. These cuckoos, among the few birds that regularly eat hairy caterpillars, are the most important avian predators of the hairy tent caterpillars. In *Birds of Massachusetts*, Edward Howe Forbush wrote of the black-billed cuckoo, "In seasons when caterpillars . . . are abundant, cuckoos usually become common in the infested localities. They follow the caterpillars, and where such food is plentiful the size of their broods seems to increase." He also reported that an observer in Indiana noted that yellow-billed cuckoos destroyed "every tent caterpillar in a badly infested orchard" and tore up all of their nests in half a day.

You don't have to take my word for this story. The role of telephone wires is based on my own observations in Connecticut, but the rest of the story can be pieced together from Forbush's book; another book on birds, which includes the life histories of cuckoos, by Arthur Cleveland Bent; Mary Willson's seminal volume on reproduction by plants; and Terrance Fitzgerald's recent and very useful book that brings together the scattered information on tent caterpillars, and on which I have relied heavily in the preparation of this chapter.

❀ ❀ ❀   There are two kinds of tent caterpillars, both highly gregarious, in eastern North America: the eastern tent caterpillar (*Malacosoma americanum*), whose big, white, silken tents are a familiar sight in wild black cherry trees along country roadsides, and the forest tent caterpillar (*Malacosoma disstria*), which, contrary to its name, does not build a tent. The latter has been called the "forest armyworm," because it sometimes migrates through the forest by millions as it defoliates trees: aspens and poplars, which are its favorites, but also oaks, maples, and many others. According to Fitzgerald, the eastern tent caterpillar has the more complex society and is, indeed, the most socially complex of all caterpillars.

❀ ❀ ❀   Eastern tent caterpillars have only one generation per year. They survive the winter as diapausing eggs in masses attached to thin twigs of their host plants and are often easy to find on wild black cherry trees. These masses, which may be as much as half an inch in diameter and three quarters of an inch long, more or less encircle the twig like a collar. Each mass contains about 300 eggs, is laid by one female, and is the only egg mass that she will produce during her brief lifetime. The eggs are packed together about as tightly as possible, and the entire mass is covered with a solidified, shiny, shellac-like secretion that may give some protection against parasites, but serves mainly to keep the eggs from drying out by absorbing and holding moisture from the air. An eastern tent caterpillar spends 9 months of its life as an egg, and like so many other diapausing insects, the egg will not hatch until after it has experienced a long period of cold weather.

By early spring, the eggs have undergone a sufficiently long period of chilling, and the tiny caterpillars chew their way out of the shells of their eggs just as the buds of their host plants are bursting to reveal the tiny, developing leaves of spring. At first, the newly hatched caterpillars aggregate on the egg mass, but eventually they move off to find leaves to eat and a site for the communal tentlike shelter that they will very soon build in one of the main forks of the tree. Sometimes caterpillars from more than one egg mass, offspring of different mothers, join forces to build a large tent that houses them all.

The shelters of eastern tent caterpillars are sometimes confused with those of fall webworms, but it is easy to tell the difference. Fall web-worm shelters are built at the ends of branches and surround the leaves on which the caterpillars feed. Eastern tent caterpillars enclose no leaves in their shelters and, unlike fall webworm caterpillars, must leave their shelter to find leaves to eat. When the tent caterpillars have grown to their maximum size, in late June or July, they stop feeding and usually leave the trees to seek a protected site, often in a crevice or under an overhang, in which to spin a sturdy silken cocoon. After completing the cocoon that encloses it, the caterpillar sheds its skin to become a pupa. During the next 2 or 3 weeks the adult develops within the pupa. When the adult becomes fully formed, it sheds the pupal skin, and, after se-creting a substance that loosens the silk at the head end of the cocoon, forces its way out and expands its wings. Females emit a sex pheromone that attracts males. After mating the female lays her single egg mass. Since she does not feed, she usually dies in less than a day.

For about half of the day, eastern tent caterpillars shelter in their tent or bask in the sun on its outer walls, but from time to time they must leave the relative safety of the tent to find food. As Fitzgerald, Tim Casey, and Barbara Joos found, they have three bouts of feeding each day: one whose midpoint is 6:00 A.M. (EST) that, on average, lasts for more than 5 hours; another whose midpoint is 1:00 P.M. that lasts for about 2.5 hours; and a third whose midpoint is 8:00 P.M. that lasts for about 4.5 hours.

The caterpillars do not dash boldly from tent to leafy branch. They seem to be timid and reluctant to leave their shelter. It seems apparent that, as we so often see in animal behavior, evolution has instilled in them two conflicting urges: hunger with its concomitant drive to venture out to where the leaves are and the innate fear of exposing themselves to danger by leaving their shelter. At first, the caterpillars gather on the outside walls of the tent, milling around for about an hour as they lay down strands of silk. Eventually, a few individuals, probably hungrier than the rest, lead the way and the caterpillars that have gathered on the walls of the nest move en masse onto branches that lead to sprays of leaves.

As the foraging caterpillars move away from the tent, Fitzgerald noted, they mark their path with a pheromone by repeatedly brushing the end of the abdomen against whatever surface they are crawling on. The pheromone trail makes it possible for these dim-sighted creatures to follow each other without touching, serves as a guide for latecomers, and leads the caterpillars back to the nest after they have fed. This pheromone has been isolated and chemically identified. Researchers collect the natural pheromone for chemical analysis by wiping the caterpillar's hind end with a piece of filter paper. This isn't what it sounds like. The creature's anus is well removed from the area where the pheromone is secreted. Fitzgerald reported that the caterpillars are very sensitive to the pheromone, able to follow a trail marked with as little as 1 trillionth of a gram of chemically pure artificial pheromone per millimeter of trail. (A gram is about 35 hundredths of an ounce, and about 25 millimeters make an inch.) There is no doubt that the caterpillars are guided only or mainly by the pheromone. In an experiment, a procession of them followed every curve and turn of an

S-shaped trail of the pheromone that an experimenter had drawn on a large card.

When they have finished feeding, the caterpillars follow the outward exploratory trail back to the tent and may superimpose on this trail a recruitment pheromone that is chemically different from the pheromone that marks the outward trail. Doubly-marked trails, which lead to known and favorable feeding sites, are more attractive and more likely to be followed by food-seeking caterpillars than singly-marked exploratory trails. Depending upon the quality of their experience at the feeding site, eastern tent caterpillars may or may not advertise a feeding site to their tentmates by laying down a recruitment pheromone. If, for example, the site was overly crowded, the returning caterpillars usually do not apply recruitment pheromone. Furthermore, in an experiment described by Fitzgerald, caterpillars allowed to eat young nutritious leaves recruited significantly more food-seeking caterpillars to their trail than did those that were forced to eat less nutritious older leaves.

Like almost all leaf-feeding caterpillars and many other insects, eastern tent caterpillars feed voraciously and put on a lot of weight in a short time. So ravenous are they that the caterpillars in a single colony often strip all of the leaves from a black cherry sapling by the time they approach their full growth and must move on to find more food. Fitzgerald's book and an article by Douglas Futuyma and Steve Wasserman give us a picture of how tent caterpillars grow and how they budget the leaves they eat. A caterpillar weighs only a fraction of a milligram when it hatches from the egg, and in about 7 weeks it will become a full-grown caterpillar that weighs close to 900 milligrams (about 0.3 of an ounce), increasing its weight by a factor of from 2,000 to 3,000. It will, of course, subsequently lose considerable weight as it passes through the nonfeeding pupal stage and metamorphoses to the nonfeeding adult stage.

During those 7 weeks it eats over 2,400 milligrams of leaves (well over 0.8 of an ounce). Of the total amount eaten, about 68 percent will be expelled as feces that consist mostly of indigestible cellulose and lignin. (The feces are not wasted; they are a resource for scavenging insects.) Thus, only 32 percent of the weight of the leaves eaten is actually assimilated by the caterpillar. Of the total amount eaten 18 percent is ex-

pended as energy to support activity and the processes of the body; 3 percent is expended for the production of silk; and, finally, 11 percent of what is eaten is incorporated as an increase in body weight.

⚙ ⚙ ⚙   No one knows exactly how hundreds of eastern tent caterpillars that have just hatched from an egg mass reach a consensus and decide on the site where they will join forces to build their tent. The tiny caterpillars crawl, on average, about 5 feet away from their egg mass before settling on a site. They almost always pick one that will give the tent favorable exposure to the sun and that is roomy enough to allow for future expansion of the tent as the caterpillars grow. A new 2-inch-long tent may ultimately be enlarged to a length of 2 feet or more.

The chosen site will be in the crotch of a major fork of the tree, often where a main branch meets the trunk. Such a site allows, within the limits of the branches that support it, ample room for expanding the tent upward and out to the sides. Since the eggs hatch in very early spring, late March in Washington D.C., the caterpillars are very likely to be exposed to low temperatures, sometimes so cold that they are unable to move. Consequently, it is important that the tent be oriented so as to receive as much warmth as possible from the sun. When first building or later enlarging the tent, the silk-spinning caterpillars concentrate on its southeastern wall, making it the widest wall so as to have maximum exposure to the sun.

The shape of the tent is largely determined by the position of the branches of the fork in which it is built, but it is always tentlike in shape and always has a discrete entrance hole, usually at the top. The tent consists of many layers of silk with wide gaps between them, layers that are formed as the caterpillars add silk to the outside of the tent. Practically all tent construction occurs when the caterpillars are gathered as a group on the outer walls of the tent before they move out to feed. These communal bouts of spinning last for about an hour, but any individual caterpillar will probably spin silk for only 15 to 30 minutes.

As Fitzgerald explained, the tent serves several functions. As we will see in detail in the next chapter, probably the most important is to modify the microclimate in which the caterpillars live. The silken fabric of the tent also provides a secure foothold for the caterpillars—particularly important on cold days in early spring when their body temperatures fall so low that they are unable to move about and can barely manage to hold onto their places in the tent. The walls of the tent are also the communications center of the colony. It is there that all of the pheromone trails that lead to feeding sites converge, and it is there that the caterpillars meet to begin their communal foraging forays.

The importance of the tent in protecting caterpillars against parasites and predators is not completely understood. As a group, the North American species of tent caterpillars are attacked by 66 birds and 170 insects, some of the latter predators but most of them parasites. Perhaps the tent somewhat lessens the depredations of these known attackers, and it presumably altogether foils other potential attackers. Some researchers contend that the protection offered by the tent is insignificant, a view based on experiments with tent-building Mexican and Japanese relatives of our eastern tent caterpillar. In experiments done with these insects in Mexico and Japan, fewer caterpillars were killed in tents covered with fine screening that excluded parasites and predators than in tents that were left uncovered. These experiments suggest that the caterpillars' tents may not give complete protection against parasites and predators, but they do not show that the tent gives no protection or that whatever protection a tent does offer is unimportant. Even a very small degree of protection could mean the difference between the death of a whole family of caterpillars, the only offspring of one mother, and the survival of just a few to pass on their mother's genes to future generations.

Tent caterpillars have group behaviors other than tent building that tend to protect them against predators and parasites. For example, western tent caterpillars gathered on the outside of their tent violently thrash the anterior parts of their bodies from side to side at the approach of a flying parasite, including a certain parasitic tachinid fly that glues a few of her eggs to the heads of caterpillars, the only place where they will be safe from the caterpillars' menacing jaws. The maggots that hatch from the eggs burrow into the caterpillar's body and ultimately kill it. Thrashing starts with a few individuals but usually spreads to all or most members of the group. It has been seen to discourage parasites.

Judith Myers and James Smith found that the caterpillars are alerted to the presence of parasites by the buzzing noise of their flight. When Myers and Smith played a tape recording of a tachinid fly's buzz near a nest, the tent caterpillars on its surface responded by vigorously thrashing their bodies. Myers and Smith also showed that the range of sound frequencies perceived by the caterpillars neatly brackets the frequency range of the fly's buzzing.

Like many caterpillars and other insects, eastern tent caterpillars regurgitate when they are disturbed by some creature such as a parasite or an insect-eating predator. The regurgitant of eastern tent caterpillars, as Steven Peterson and his coworkers discovered, contains deadly hydrogen cyanide and the volatile and repellent benzaldehyde, both compounds derived from the cherry leaves that the caterpillars eat. In laboratory tests, only a few ants—potentially important predators of tent caterpillars—were repelled by the dilute vapor of hydrogen cyanide alone but most of those tested were repelled by the vapor of the benzaldehyde that accompanies the hydrogen cyanide. The young cherry leaves that eastern tent caterpillars prefer are not only more nutritious than old leaves but also contain more cyanide and benzaldehyde. Stevenson and his coworkers found, as is to be expected, that in laboratory studies the regurgitant of caterpillars fed young leaves is more repellent to ants than is the regurgitant of caterpillars fed old leaves. As Fitzgerald wrote, little else is known about this behavior or its effect on predators. This regurgitant obviously does not deter cuckoos and some other birds from eating tent caterpillars, but we don't know its effect on other insect-eating birds, other vertebrates, and arthropods—insects and their multilegged relatives. An experiment that might answer this question—one which, as far as I know, has not yet been tried—is to determine the reaction of insect-eating predators to other species of caterpillars that are known to be palatable but have been smeared with the regurgitant of tent caterpillars.

⚙ ⚙ ⚙   In addition to the several species of tent caterpillars, quite a few other caterpillars are gregarious. Many of them live "shoulder to shoulder" in closely packed groups and are obviously attracted to each other, and some lay down pheromone trails that are followed by group members. But few interact in ways that approach the complexity of the interactions between the members of a colony of eastern tent caterpillars.

Members of some gregarious species other than tent caterpillars do, however, build communal shelters of silk or a mixture of silk and leaves, and in some species, individuals from unrelated families—different egg masses—even cooperate with each other, as eastern tent caterpillars sometimes do, in building a single communal tent that will house all of them. Among the caterpillars that cooperate with unrelated families are those of a butterfly, the Baltimore checkerspot, and those of three moths, an ermine moth, the ugly nest moth, and the coconut moth.

Although multifamily caterpillar societies are relatively simple in organization, in the composition of their colonies, in one way they resemble human societies more than do the more complex and organized societies of the most social of the insects, the termites, ants, and certain wasps and bees. Like a human society, these caterpillar colonies include unrelated individuals. By contrast, with few exceptions, a colony of termites, ants, wasps, or bees is a single family unit. All the workers in a honey bee colony, for example, are sisters or half sisters, all offspring of the same queen, which mated with several males and is the only member of the colony that is capable of laying eggs that will produce the females that will become workers or new queens.

There are, however, some exceptions among these more complex (eusocial) insect societies. Some colonies are founded by two or more queens. In some species, these foundresses are sisters, and thus the colony is not a multifamily unit in the same sense as are some caterpillar colonies. But in other eusocial species, including some wasps, ants, and bees other than honey bees, colonies are founded by several unrelated females. This, according to Steven Rissing and Gregory Pollock, is what happens in certain species of leafcutter ants. Rissing and several coworkers used genetic tests to show that the cofoundresses of these colonies are no more closely related than are a randomly selected group of queens.

❂ ❂ ❂   Insects, because they are so small, are more sensitive to unfavorable weather conditions than are larger animals and have, conse-

quently, evolved a variety of ways to ameliorate their microclimates. They cannot, of course, stop winter from coming or keep the summer cool, but like the eastern tent caterpillars with their silk tent, many of them—especially if they live in groups—display an often impressive ability to control the temperature and humidity of their immediate surroundings. In the next chapter we will look more closely at how tent caterpillars control their environment and see how other insects and their relatives make use of their surroundings to improve their lot in life.

Controlling the Climate

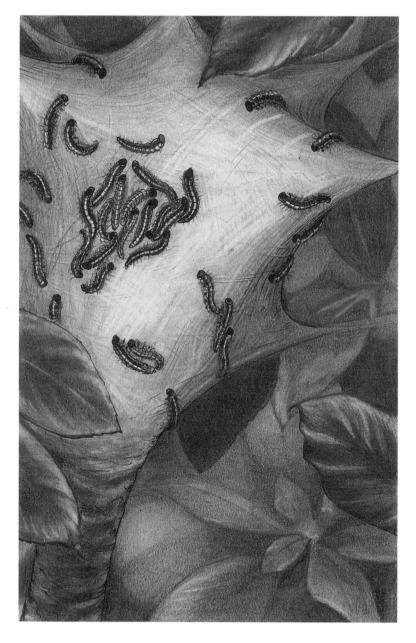

*Tent caterpillars basking in the sun as they sit
on the outer wall of their nest*

❀ ❀ ❀ How crucial the ability to affect the microclimate is for insects was made startlingly clear to William Wellington when he observed the sudden, life-saving escape of a large group of western tent caterpillars (*Malacosoma californicum*) to their silken shelter at the approach of a cold front. The caterpillars had been peacefully munching on the leaves of a red alder tree, when suddenly—in the space of just a few seconds—all of them turned and rushed back to their tent. In less than a minute, all the caterpillars of the colony were inside the tent. As Wellington put it, it had been a veritable stampede compared to the leisurely nose-to-tail procession in which these tent caterpillars usually return to their shelter after they have finished feeding. Furthermore, the caterpillars usually gather for a while on the outer surface of the tent before entering. This time they sped into the tent without dawdling for a moment.

Wellington and his assistants had placed sensors that are responsive to slight changes in temperature and moisture level in the area frequented by the caterpillars: in the air, on leaves, on twigs, and on the tent itself. The recordings from these sensors showed no changes prior to the caterpillars' stampede, and Wellington was about to conclude that a change in the weather had not precipitated their hurried departure. But a few moments later, all of the sensors showed a sharp drop in temperature, and soon thereafter it became obvious that a cold front had arrived. How the caterpillars detected this change in the weather so early remains a mystery, although they may have sensed a change in air pressure, which Wellington had not measured. Nevertheless, they had gained the safety of their tent well before the cold front arrived. There they were protected from the ensuing wind and rain and were snugly warm, because the walls of the tent slowed the loss of heat from its interior.

Western tent caterpillars, like eastern tent caterpillars, are active in early spring when the weather is variable and cold days are interspersed with warm days. Accordingly, both species have evolved be-

haviors that enable them to cope with the vagaries of the weather. Foremost among these behaviors is the communal construction of a tent by the members of a colony. The tent acts not only as a shelter from wind and rain, but also as a greenhouse, just as does one made of glass, by allowing the heat-producing rays of the sun to enter more easily than heat rays (infrared rays) can escape. The tent is remarkably efficient. When the outside air temperature was 46°F, the temperature inside an empty tent of the eastern tent caterpillar was as high as 70°F in the center of the structure, although somewhat lower at other locations. The silken walls of a tent also slow down the loss of moisture, thus protecting the caterpillars against desiccation when the outside air is too dry for them. This effect is nicely illustrated by the behavior of western tent caterpillars, as Terrance Fitzgerald has pointed out. In the humid climate of the northern Pacific coast, caterpillars of this species commonly gather to rest on the outsides of their tents. But in the exceedingly dry climate of the Mojave Desert, they usually rest inside their tents, presumably to take advantage of its moisture conserving quality.

Eastern tent caterpillars often bask in the sun on the outside wall of the tent to warm themselves on cool days. Basking by tent caterpillars occurs mainly in the morning, when the caterpillars return from their first feeding bout of the day. Raising their body temperature speeds up the digestion of the leaves they ate that morning. They gather in a tight mass, body pressed to body, on the outside surface of the wall on the sunny side of the tent. Such a dense mass of the dark caterpillars is a remarkably efficient absorber of solar heat, and individuals within the aggregation are further protected from the heat-dissipating wind by the hairy bodies of their surrounding nestmates.

A closely packed group of dead tent caterpillars placed in spring on the sunny southeast wall of a tent had, as Fitzgerald noted, body temperatures that averaged 75°F *above* the temperature of the surrounding air. Even if the air temperature was low, close to freezing, that would raise a live caterpillar's body temperature to a lethal 107°F. But a live caterpillar would, of course, cool itself by moving into the shade before its temperature reached the lethal level.

If the great amount of heat absorbed by the dark caterpillars surprises you, consider the simple experiment that physics teachers sometimes do to demonstrate the difference in the ability of light- and dark-colored objects to absorb heat. On a cold but sunny day in winter, two plates of sheet metal are placed on the surface of the snow so that the sun shines directly upon them. One plate has been polished and reflects the sun's rays. The other has been painted matte black so that it absorbs the sun's rays. After a few hours the black plate will have melted the snow beneath it and sunk well below the surface. But the shiny plate will still be sitting on top of the snow.

⚙ ⚙ ⚙   Being small has its advantages, but it also has its disadvantages. Because of their small size, insects are far more susceptible to unfavorable humidities and temperatures than are larger animals. Moisture and heat are gained or lost according to an animal's surface area—the larger the surface area the greater the gain or loss. And the surface area of a small animal is greater in proportion to its volume or weight than is that of a larger animal.

This is so because the volume of any object increases as the cube of its length, while its surface area increases only as the square of its length. Let's imagine as an example a cubical block that is 3 inches long on each side. The volume of the block is 3 cubed ($3 \times 3 \times 3 = 27$ cubic inches). The area of each of its six sides is 3 squared, and its total area is the area of a side multiplied by 6 ($3 \times 3 = 9$; $6 \times 9 = 54$ square inches). If we do the same arithmetic with a cubical block that is only 2 inches long on each side, we will find that its volume is 8 cubic inches and that its surface area is 24 square inches. By subtracting only 1 inch from the dimensions of the cube we have increased its volume-to-surface area ratio by a factor of 3 (24/8) from a factor of 2 (54/27).

Although their dimensions are much harder to measure, it works the same way with the body of an insect or any other animal. Obviously, a small animal will lose heat or moisture more rapidly than a larger one. This is why hummingbirds, the smallest of birds, must eat proportionally more than larger animals and become torpid at night and lower their body temperature to reduce heat loss. And keeping up with the great loss of heat from their tiny bodies is one of the reasons why shrews, the smallest of all mammals, must eat as much as two or three times their own body weight every day.

Although insects that live in groups have an advantage, even lone in-

sects can, as Vincent Wigglesworth explained, control their body temperature and the loss of water from their bodies to a considerable degree. An insect that is too cold can move into the sun. Many insects that bask in the sun position their bodies so as to expose the maximum surface area to the sun's rays. If you go outside just after dawn on a cool morning you can find grasshoppers sitting at the tops of plants with their broad sides facing the rays of the sun. Similarly, you can find butterflies perched in an exposed position with their wings spread wide and held flat out to the sides so as to gain maximum exposure to and thus maximum benefit from the sun's rays. In an experiment described by Wigglesworth, a desert locust that sat with the broad side of its body toward the sun had a body temperature that was about 60°F above the air temperature, while one that sat facing the sun, and thus exposing less of its surface area to the sun, had a body temperature that was only about 43°F higher than the air temperature. Lone insects can also raise their body temperature by becoming active, particularly by fluttering their wings or shivering their wing muscles. Heat is a by-product of the metabolism that provides the energy for these activities. Some sphinx moths, which are active during the cool of the night, cannot fly unless their body temperature is at or above 90°F. Even on a cool night, they can raise their body temperature well above 90°F by fluttering their wings as they sit. If a lone insect is too hot it can do little but move to a cooler place—sometimes by moving into the shade of a plant or sometimes by burrowing down to a cooler layer of the soil. But as you will read below, the social honey bees and termites have amazingly efficient ways of cooling their colonies.

Many terrestrial insects need to conserve the water in their bodies. Most of them do not have access to drinking water, although some drink droplets of dew that form on plants in the early morning. But the only moisture that most insects get is the water content of their food, which may be as much as 92 percent in the cabbage leaves that a caterpillar eats or as little as 12 percent or less in the dry beans that the beetles known as bean weevils eat.

Insects can decrease the evaporation of water from their bodies by moving to a cooler place, and all of them have physiological and anatomical characteristics that help conserve the water they contain. The insect's outer skin, or cuticle, not only serves as its skeleton, but is also a waterproof covering that largely prevents evaporation from the body. The cuticle is relatively thick, but its two outermost layers are almost microscopically thin. The inner one is a layer of wax that blocks the evaporation of water through the cuticle, and the outer one, called the cement layer, protects the soft wax layer from being abraded. Experiments in which the wax layer was partially removed show that it is an anatomical feature essential to the conservation of water. (Insects with the wax layer intact survived, but those whose wax layer had been abraded eventually died.)

The physiological mechanism that conserves water is fascinating. The wastes from the metabolism of nitrogen—the characteristic component of proteins—include ammonia, urea, and uric acid, all of which are toxic and must be removed from the body in a timely manner. Mammals, including humans, eliminate nitrogenous waste from the body as urea, which, since it is very toxic and highly soluble, must be flushed out of the body with large quantities of water. But insects, like birds, conserve water by excreting uric acid, which is less toxic and much less soluble, and can therefore be eliminated from the body with relatively little water. Both birds and insects eliminate their "urine" with the feces. The uric acid in insect feces is invisible because it is thoroughly mixed with the wastes of digestion, but in birds it is easily visible as a large, almost dry, white blob at the end of the fecal pellet.

⊛ ⊛ ⊛   Some relatives of the insects have striking group behaviors that help them conserve moisture. One species of the harvestmen, or daddy-longlegs, known as shepherd spiders in England, forms very large and amazingly dense aggregations that are without doubt a means of retaining moisture transpired by its members. In an article published in 1954, Helmuth Wagner of the Überseemuseum (Overseas Museum) in Bremen, Germany, described an aggregation of these relatives of the spiders that he found in early April in a dry, treeless, mountainous area in the state of Jalisco in Mexico. No rain had fallen in April and only about a half inch had fallen the previous March and February. As Wagner wrote, it was so dry that no dew formed in the mornings.

These harvestmen, about 70,000 of them, were tightly packed into the crotch where the three main branches of a 15-foot-tall candelabra cactus joined. The aggregation, so dense that its members were tightly pressed together, was almost 2 feet high and had a circumference of almost 36 inches. The harvestmen at the periphery of the aggregation lay on their backs, with their long, threadlike legs extending straight up above them. The thousands of legs sticking out from the surface of the aggregation obscured the bodies of the harvestmen and made the aggregation look like a thickly-haired pelt. This "pelt" helped slow the loss of moisture from the surface of the mass of harvestmen by diverting the wind.

Such an aggregation of creatures is, of course, a strong temptation to almost any hungry animal that comes along—ranging from ants to reptiles, birds, and small mammals. But the harvestmen are not attacked because they have an effective chemical defense. Wagner reported that those in the aggregations on candelabra cacti secreted a noxiously odoriferous substance from a pair of glands at the front end of the body, and that its odor was perceptible from more than 15 feet away. Perhaps the legs of the harvestmen at the periphery of the aggregation aid in the dissemination of this substance. In 1971, Thomas Eisner and three coauthors reported that a different species of harvestman from Texas chases off marauding ants by using the front legs to dab them with a noxious secretion. This harvestman dips its two front legs into the secretion from glands between the bases of the first and second pairs of legs, and then dilutes it with regurgitant before smearing it onto the ants.

❁ ❁ ❁   Harvestmen are not the only creatures that form aggregations to conserve moisture. Any tightly packed aggregation of insects, including the ladybird beetles you met earlier, will probably decrease the rate at which its individual members lose water. David Denlinger, an entomologist at Ohio State University, and his coworkers have been observing a huge aggregation of bright red and presumably chemically protected beetles that forms on the same tree each year, and have measured the water loss of lone individuals and individuals clumped in groups of different sizes. Denlinger described this aggregation of diapausing beetles in the *American Entomologist:*

From a distance it looks just like another tree. There are plenty of other specimens of [this] palm . . . on Barro Colorado Island, Panama, and this one is not a standout. It is not on a conspicuous promontory nor does it project a profile that dominates the landscape. It is just one tree among many on the sloping terrain of this remnant of Panama's rain forest. Yet, it is to the base of this tree that thousands of adult handsome fungus beetles . . . descend each July and August. By the time they all arrive, their number may exceed 70,000. They stay, clumped in this huge aggregation, throughout the remainder of the rainy season, through the four-month dry season that begins in early January, and then disperse when the rains return in late April. A few months later a new generation of adult beetles comes back to the very same tree. This pilgrimage of beetles has been going on for at least thirteen years.

Denlinger and his colleagues J. L. Yoder and Henk Wolda measured the effect of group size on water conservation by these handsome fungus beetles. Lone beetles and groups of 25, 100, or 250 were placed in containers with a relative humidity of 0 percent. The beetles soon gathered together in clumps. The weight loss, essentially equivalent to water loss, of the lone beetles and a sample of 15 marked beetles from each group was measured once every hour. The lone beetles lost water the most rapidly, at a rate of 0.48 percent of their weight per hour. Beetles from the group of 250 lost water the most slowly, at a rate of only 0.22 percent of their weight per hour. Beetles in the smaller groups lost weight at intermediate rates.

With Seija Tanaka and Henk Wolda, Denlinger found that group size also affects the metabolic rate of handsome fungus beetles. The smaller the group and the drier the air, the higher the metabolic rate. Denlinger and his colleagues did not suggest a reason for these increases in metabolic rate, but they did find that these beetles are very intolerant of low humidity. At about 75 percent relative humidity, beetles in clusters of 25 to 30 individuals all died within 2 weeks. But groups of the same size kept at 100 percent relative humidity survived for many months.

I suggest that the metabolic rate increases with the dryness of the air—which is, of course, greatest in small groups—in order to alleviate the beetles' "thirst." The metabolism of a sugar or a fat yields water as a

by-product, often referred to as "metabolic water." For example, the complete metabolism of an ounce of fat yields slightly more than an ounce of water. Some insects, including many of those that live in stored grain, obtain little water from the dry food they eat, and since they have no access to drinking water, depend largely upon metabolic water. Perhaps the handsome fungus beetles on Barro Colorado Island use metabolic water to keep themselves alive when they are dry. Using their store of fat to produce metabolic water shortens the period during which they can survive without eating, but this may be a worthwhile trade-off, because desiccation greatly hastens their death, and metabolic water just might keep them alive until it rains.

⚙ ⚙ ⚙   Termites, social insects which live in colonies that, in some species, contain 2 million individuals or more, are often incorrectly referred to as "white ants." But they are certainly not ants. Termites, unlike ants, have gradual metamorphosis with only three life stages: egg, nymph, and adult. Ants and the other social members of their order, certain bees and wasps, have complete metamorphosis, four life stages: egg, larva, pupa, and adult. The worker and soldier castes of social ants, bees, and wasps consist of only females, all daughters of a single queen that mated soon after she matured and thereafter never mated again. The worker and soldier castes of termites consist of both males and females, and the queen lives permanently with a male consort.

Since termites are small and soft-bodied, they easily become desiccated and must live in moist places with a high relative humidity. They do best when the relative humidity in their nest is above 96 percent and the temperature is fairly high, an optimum of about 79°F for temperate zone species and about 86°F for tropical species. Subterranean termites, the familiar and destructive species that occurs throughout the eastern United States, attain these conditions by nesting in moist soil that is in contact with wood, their only food. The surrounding soil keeps the nest moist and tends to keep the temperature at a more or less favorable level. When it is cold in winter, subterranean termites move to burrows below the frost line.

Some tropical termites are far more sophisticated, constructing huge above-ground nests with built-in "air conditioning" that keeps the nest moist; warm or cool, as the case may be; and well supplied with oxygen. Among the most architecturally advanced of these termites is an African species, *Macrotermes natalensis*. Martin Lüscher described the mounds of this fungus-growing species as being as much as 16 feet tall, 16 feet in diameter at their base, and with a cement-like wall of soil mixed with termite saliva that is from 16 to 23 inches thick. The thick and dense wall of the mound insulates the interior microclimate from the variations in humidity and temperature of the outside atmosphere. Several narrow and relatively thin-walled ridges on the outside of the mound extend from near its base almost to its top.

According to Lüscher, a medium-sized nest of *Macrotermes* has a population of about 2 million individuals. The metabolism of so many termites and of the fungus that they grow in their gardens as food helps keep the interior of the nest warm and supplies some moisture to the air in the nest. The termites saturate the atmosphere of the nest, bringing it to about 100 percent relative humidity, by carrying water up from the soil.

But how is this well-insulated nest ventilated? Its many occupants require over 250 quarts of oxygen (almost 1,270 quarts of air) per day. How can so much oxygen diffuse through the thick walls of the mound? Even the pores in the wall are filled with water, which almost stops the diffusion of gases. The answer lies in the construction of the nest. The interior consists of a large central core in which the fungus is grown; below it is a "cellar" of empty space; above it is an "attic" of empty space; and within the ridges on the outer wall of the nest, there are many small tunnels that connect the cellar and the attic. The warm air in the fungus gardens rises through the nest up to the attic by convection—in other words, because warm air rises. From the attic, the air passes into the tunnels in the ridges and flows back down to the cellar. Gases, mainly oxygen coming in and carbon dioxide going out, easily diffuse into or out of the ridges, since their walls are thin and their surface area is large because they protrude far out from the wall of the mound. Thus air that flows down into the cellar through the ridges has been cooled or warmed according to the outside temperature, is relatively rich in oxygen, and has lost much of its carbon dioxide. It supplies the nest's in-

habitants with fresh oxygen as it rises through the fungus-growing area back up to the attic.

⚙ ⚙ ⚙   Honey bees can more precisely and consistently control the microclimate in their nest than can any other social insect—probably more so than any other insect of any sort. Even on the hottest days of summer, they can keep their nests comfortably cool by means of an amazingly efficient method of air conditioning. When the weather is cold, they huddle in their nest and keep themselves warm enough to remain active throughout winter. An individual honey bee is as "cold-blooded" as any other insect: it can raise its body temperature only through voluntary muscular activity. But a colony of thousands of honey bees acting together, a "superorganism" as Edward O. Wilson calls it, is as "warm-blooded" as a bird or a mammal. Unlike other temperate zone insects, honey bees do not survive winter by becoming inactive and entering diapause. They will even fly out of the nest on warm days in winter.

As Wilson described it, a swarm of bees chooses a site for their nest, often a manmade hive or a hollow in a tree, that will tightly enclose the colony, and then seals all crevices and openings except the entrance with plant gums. The bees control the microclimate within this tight enclosure with astonishing precision. From early spring to late fall, the period during which the colony raises its brood, eggs, larvae, and pupae, the temperature within the nest is almost constantly kept between 94° and 96°F, even when the outside temperature is near freezing. The ability of honey bees to cool their nests is even more impressive. For example, when Martin Lindauer placed a hive in the sun on a black lava field near Salerno in southern Italy, the bees kept the temperature in the hive at 95°F although the outside temperature above the lava at the level of the hive was an astonishing 158°F.

How do honey bees accomplish these amazing feats? The short answer is that they can produce heat by vibrating their wing muscles without moving their wings, and that they can lower the temperature by evaporating water within the nest.

When the weather is cool during their reproductive season, workers crowd more or less closely around the brood and generate enough heat to keep it at the optimal temperature of 95°F. Keeping warm during the really cold days of winter is a more complex process. When the air temperature in the nest falls to about 55°F, the workers of the colony, thousands of them, form a tight cluster that surrounds the queen and some combs filled with honey. The workers at the periphery of the cluster are packed "shoulder to shoulder," forming a living blanket that is two bees deep and insulates the rest of the cluster. The inner bees are less tightly packed and have enough room to move around. They keep the cluster warm by eating honey and converting its calories to heat by vibrating their wing muscles without moving their wings. From time to time the workers shift positions so that even the bees in the peripheral "blanket" can fill up on honey.

When the temperature in the nest climbs above the optimum in the summer, the workers cool it down by behaviors that become progressively more complex and effective as the outside temperature rises and becomes more threatening. When only a little cooling is required, workers fan their wings to circulate the air in the nest. When more is needed, they spread thin layers of water on the combs and let them evaporate with no further assistance. If circulating the air or passive evaporation is not enough, the workers increase evaporative cooling. They collect water, bring it into the nest, spread it thinly on the combs, and then fan their wings to hasten evaporation.

⚙ ⚙ ⚙   Although we don't usually think of them that way, the warmth of the sun and the moisture in the air are essential resources for almost all insects. Although some insects can compensate for unfavorable weather conditions by more or less controlling the temperature and humidity of the microclimates in which they live, they and virtually all other insects time their life cycles so that their active and growing phases coincide with the time of year that offers them favorable climatic conditions. Most temperate zone insects diapause in winter and are active only in spring and/or summer, when it is warm and food is plentiful. A few, however, have adapted to the cold of winter, when predators are few and there is a minimum of competition for what little food is available. In the tropics, many insects are active during the rainy season and enter diapause to avoid the rigors of the dry season. Just as in-

sects must coincide with the climatic conditions to which evolution has adapted them, they must also coincide closely in time and space with other necessary resources, such as food. Since many plant-feeding insects are host-specific—they will feed on only a few closely related plants—they must coincide in time and space with their particular host plant when it is in a growth stage that is favorable to them. In the next chapter you will learn of the remarkable ways in which insects accomplish this.

Coinciding with Resources

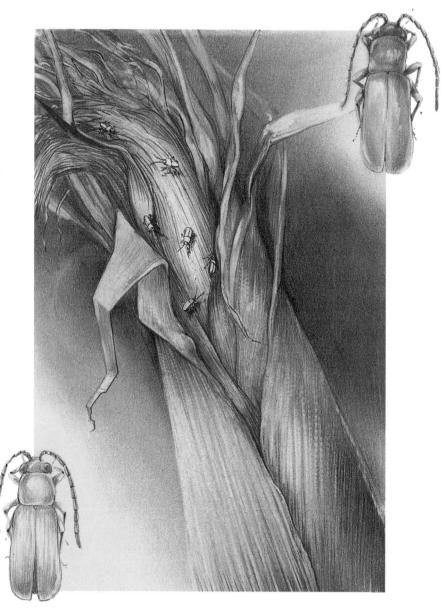

Western and northern corn rootworm beetles (enlarged at the upper right and lower left, respectively) feeding on corn silk

❁ ❁ ❁ March is the beginning of the austral (Southern Hemisphere) autumn and the beginning of the boreal (Northern Hemisphere) spring. In that month, the sandpipers known as red knots grow fat as they feed heavily on the young mussels (spat), worms, crabs, and other invertebrates that are exposed on the tidal flats of the ocean beaches of Tierra del Fuego and Patagonia. They are also completing the molt that replaces their worn flight feathers and are well on the way to replacing their drab winter body plumage with the breeding plumage, with its robin-red breast and belly, that gives them their name. In a few days, plump with the fat that will fuel the first leg of an exhausting migratory flight, the knots—probably triggered by the short days of the austral autumn—will begin a journey of almost 10,000 miles that will take them from the southern tip of South America, where they live during the austral summer, to the Canadian Arctic, where they breed during the boreal summer, prospering as they feed on the exuberant flush of invertebrate life that flourishes during that short Arctic season.

As the knots move northward, they make only a few stops to refuel, to replenish their stores of body fat by gorging on the little clams, snails, polychaete worms, and other bounty of tidal flats. They fatten up quickly. As Brian Harrington wrote, "the red knot is an amazingly efficient machine for converting marine invertebrates into forward motion through the air." One of their regular refueling stops is made during the last 2 weeks of May at Delaware Bay—actually only at certain beaches along the bay, notable among them for many years Reed's Beach on the western shore of Cape May, New Jersey. Here horseshoe crabs provide them with the abundant food, their eggs, that is essential to completing the last thousand miles of their migration. Horseshoe crabs are distant relatives of the insects. Like them they are arthropods, animals with segmented bodies and jointed legs.

Highly gregarious, the migrating knots stay close together in large flocks, a behavior that, along with providing other benefits, helps them cope with predators. At the first threat from a falcon or some other raptorial bird, they tighten their formation and, flying almost wing tip to wing tip in perfect synchrony as if they were in telepathic communication, they make swooping, zigzagging, evasive maneuvers that confound aerial predators.

Even as the red knots are migrating northward, the horseshoe crabs in Delaware Bay, probably triggered by the long days of the northern spring, are beginning to move from the deep waters of the bay toward the beaches on which they will spawn, among them Reed's Beach. The knots and the horseshoe crabs are destined to meet. The spawning of the horseshoe crabs peaks during the second half of May and the beginning of June, coinciding with the arrival of the tired and hungry knots. The eggs of these marine arthropods are an essential food for the knots, and the population of these wonderful birds would without doubt plummet if the horseshoe crabs of Delaware Bay became scarce or were extirpated. Where else would so many knots get the energy to complete the last leg of the migration to their Arctic breeding grounds if they could not first fatten up on horseshoe crab eggs? Although the knots feed voraciously, they have little or no effect on the horseshoe crab population, because the crabs lay a huge surplus of eggs, far more than are required to perpetuate their population, many of the eggs destined to become food for shorebirds and other creatures.

As Peter Ward relates in *On Methusaleh's Trail,* the four species of horseshoe crabs that exist today, three on the east coast of Asia and one on the east coast of North America, are "living fossils," the almost unchanged remnants of an ancient group of many species that flourished about 360 million years ago during the Carboniferous Period, the age of coal formation. The name horseshoe *crab* is misleading. These creatures are certainly not crabs, in no way related to them or to shrimps, lobsters, crayfish, or other allies of the true crabs. The horseshoe crabs are a group by themselves, arthropods distantly related to the extinct trilobites and the scorpions, spiders, and other arachnids, and, of course, even more distantly related to the insects.

The great majority of horseshoe crabs of the Atlantic coast spawn on spring tides, the two highest tides of the lunar month, which occur at the time of a new or full moon, when the sun and the moon are in alignment. Thousands of them crowd the beaches—according to Harrington, about 80,000 on just a half mile of a New Jersey beach on Delaware Bay.

A spawning run begins as males move into the shallows at the foot of a beach just as the tide begins to ebb. When the females move in later, males crawl onto their backs and hold on as the females crawl up onto the wet sand just above the lapping waves. Males usually outnumber females by better than four to one. Thus it is not unusual to see a female with a "conga line" of three or four males on her back, the first male grasping the female and the other males holding onto the male in front of them. Males that can't find a place on a female crowd around females that are already carrying males and follow them onto the beach. The female scoops out a shallow depression in the wet sand, and as Carl Shuster reports, lays in it 3,000 or more grayish-green eggs that are about the size of a capital "O" on this page. The female's consorts spill their semen over the newly laid eggs and then help her cover them with wet sand.

Some of the eggs remain safely buried in the sand, but many are uncovered and scattered by succeeding waves of horseshoe crabs as they crawl ashore to spawn. "Before long," as Harrington wrote, "billions of 'surplus' eggs lie in heaps on the shore." They are everywhere, and windrows of them are left behind by receding tides. This great bonanza of food, rich in oil and protein, attracts hordes of migrating shorebirds: not only red knots but also dunlins, semipalmated sandpipers, sanderlings, ruddy turnstones, and others, every bird eager to replenish its store of energy-rich body fat before continuing its northward migration. These surplus eggs are not just waste from the crabs' point of view. They serve the crabs by satiating the appetites of shorebirds that might otherwise probe in the sand for eggs that do have a chance of surviving.

When the red knots from Patagonia, as many as 100,000 of them, arrive at Delaware Bay, their only important feeding stop in North America, they have already flown 8,750 miles, at least 2,000 miles of that over the open Atlantic with no possibility of finding food. They desperately need to put on enough fat to see them through the remaining 1,000 miles of their journey to the Arctic, a need that is filled by the bonanza of horseshoe crab eggs on the beaches of the bay. Harrington calculated that, on the average, a red knot will double its weight in about 2 weeks by eating about 135,000 of these eggs.

The knots are currently threatened by a severe decline in the horseshoe crab population of Delaware Bay. Joan Walsh of the Cape May Bird

Observatory told me that in New Jersey there has been a 90 percent decline in the number of reproducing adult horseshoe crabs and a 99 percent decline in the number of their eggs. In 1997 the knots deserted Reed Beach in Cape May and shifted to beaches across the bay in Delaware, where horseshoe crab eggs were apparently more plentiful. Other factors are probably involved, but this decline seems to be due largely to a nonsustainable commercial harvest of the crabs. Trawlers take them from the bay by the boatload! Semi-trailer loads of them are shipped up and down the Atlantic coast and they are sold for as much as a dollar each as bait for eel pots and whelk traps, and some are shipped inland to bait traps for wild freshwater catfish. About 10 percent of the adult population is harvested each year—far too many, since it takes 9 years for a horseshoe crab to become sexually mature. New Jersey is in the process of drafting a plan to conserve the crabs.

During the spring migration, the largest concentration of sandpipers and other shorebirds that can be seen in the eastern United States gathers on the shores of Delaware Bay. This spectacle attracts multitudes of bird watchers to the area—especially to Cape May. Threats to these migrant shorebirds from industrial pollution, oil spills, and the overharvesting of horseshoe crabs are a serious concern to Cape May and surrounding towns. Bird watchers and other eco-tourists contribute over $30 million to the local economy every year.

✿ ✿ ✿   All living things must, just as do red knots and horseshoe crabs, coincide in space and time with favorable environmental conditions, members of the opposite sex, and necessary resources such as food or nesting sites. In short-lived species, this requires precise timing and synchronization. For example, the reproductive success of mayflies, many of which survive as adults for only a day, is maximized if all individuals in an area are ready to mate on the same day.

Floodwater mosquitoes are so called because, unlike other mosquitoes, they lay their eggs in crevices in dry soil that is subject to periodic flooding, such as temporary woodland pools and the flood plains of rivers. The aquatic larvae do not hatch from the eggs until after they have been inundated, and then only if certain other requirements have been met. Then all of the eggs that have accumulated hatch at the same time, coinciding with the food resources that the larvae require and ultimately giving rise to hordes of adult mosquitoes, sometimes huge swarms of them that look like clouds of smoke rising from marshes and other breeding areas.

There are many kinds of floodwater mosquitoes, but the most familiar of them—one with which you have probably had intimate contact— is *Aedes vexans*, also known as the inland floodwater mosquito but often simply referred to as *vexans* by entomologists. Where it is present, it is likely to be the most abundant and annoying of the pest mosquitoes. It occurs over much of the world: throughout the United States and southern Canada, in Europe, western Africa, eastern Asia, and on most of the Pacific islands from Japan south to Samoa.

The larvae of *vexans,* and indeed those of the other floodwater mosquitoes, are exceptionally closely attuned to the condition of their habitat, more so than many other insects. First, these mosquitoes are broadly attuned to the seasons by their winter diapause. The short days of late summer and early autumn stimulate the females to lay diapausing eggs. They lie dormant through the winter and, under normal circumstances, cannot be reactivated until after they have experienced a period of chilling. Diapause can be terminated in the laboratory by keeping the eggs on a piece of moist filter paper for a month at 40° Fahrenheit. Second, even after diapause has been terminated, the eggs do not hatch, either in the laboratory or in the field, until after they are covered with water. Finally, even if the eggs are covered with water, the larvae, which lie fully developed in the eggs, will not pop out of their shells until after they perceive a cue that tells them that bacteria and other microorganisms, the food that they will strain from the water, are present and abundant enough to satisfy their appetites.

What is the pertinent cue? The first hint to researchers was that *vexans* eggs, which hatch within minutes after being covered with water containing organic substances, do not hatch if they are covered with pure distilled water, even if they are held there for many days. Early in the 1950s, William Horsfall, a medical entomologist and mosquito expert at the University of Illinois in Urbana-Champaign, proposed the hypothesis that the cue is a decrease in the concentration of dissolved oxygen in

the water that is caused by the oxygen-consuming metabolism of a large population of bacteria and other microorganisms that are nourished by organic substances in the water.

Alfred Borg, a microbiologist, and Horsfall designed an ingenious experiment to test this hypothesis. Mosquito eggs were placed in a container of distilled water that had the usual high ambient concentration of dissolved oxygen. They did not hatch. But if the concentration of oxygen in the water was sufficiently decreased by bubbling nitrogen through the water, the mosquito larvae almost instantly popped out of their eggshells, although no bacteria or other microorganisms were in the water. In subsequent experiments, the container of eggs and distilled water was placed in a chamber with an atmosphere of nothing but nitrogen. In both cases, the amount of oxygen in the water rapidly decreased as oxygen diffused from the water into the nitrogen.

✿ ✿ ✿ If you ask farmers almost anywhere in the corn belt of the United States to name the insects that are currently most destructive to corn, they will put the corn rootworms at or near the top of their list. These small creatures are not worms at all. Rather, they are insects, the larval stages of two kinds of beetles, the northern and western corn rootworms. They live in the soil and chew on and tunnel within corn roots, often occurring in large numbers and destroying almost the entire root systems of many plants. They will eat only the roots of corn and a few other grasses, starving to death rather than feeding on a nongrass such as soybean or a native broad-leaved plant such as cone flower or prairie sunflower. The adults do not live in the soil and are not nearly so fussy about what kinds of plants they eat.

Corn (maize), as Paul Mangelsdorf wrote, evolved to become a species separate from its ancestor, which was probably a tall grass known as teosinte, as it was being domesticated and agronomically improved by native Americans, probably in the valley of Mexico City. During the 4,000 or 5,000 years that corn has been in existence, the northern and western corn rootworms have become largely dependent upon this plant and are associated with it throughout the corn belt of the United States and southern Ontario.

Before they came to feed on corn, both rootworms fed only on the roots of perennial wild grasses. An association with wild annual grasses, which live for only one season and do not survive the winter, was and continues to be an impossibility for these insects because of their seasonal history and egg-laying behavior.

In North America, the two rootworms, each of which has only one generation per year, survive the winter only in the form of diapausing eggs in the soil. The adults lay their eggs in late August or early September in the soil—but only at the base of a living grass plant, usually a corn plant today but a perennial grass in preagricultural times. The adults are dead by the onset of winter. When the eggs hatch in the spring, the tiny larvae, which can travel for only a short distance through the soil, die unless living grass roots—in the corn belt, almost always those of corn—are nearby. By early July, most of the rootworm larvae are fully grown and move away from the roots to pupate nearby in cells that they form in the soil. In July and August the adult beetles shed their pupal skins and emerge from the soil to feed, mate, and lay their eggs. Adults are willing to feed on more kinds of plants, including nongrasses, than are larvae—sometimes on leaves but preferably on pollen and flowers. Today they feed mainly on the silks and pollen of corn.

The behaviors that enabled northern and western corn rootworms to exploit wild perennial grasses predisposed them to similarly exploit corn, which is, after all, a species of grass. One obstacle in the way of the rootworms' switch from perennial wild grasses to corn is that corn is an annual rather than a perennial. But this obstacle was largely obviated by a common practice of many corn growers: planting corn in the same field year after year. Although corn is indisputably an annual, this practice makes it effectively a perennial, available to newly hatched corn rootworm larvae spring after spring.

Although the two species of rootworms now have similar biological associations with corn and have similar adverse and sometimes disastrous effects on corn, they came to be associated with this crop in different ways and in widely separated geographic areas. There is little doubt that western corn rootworms first became associated with corn in Meso-America, probably Mexico, thousands of years ago when corn was first being cultivated. It may well be that, even earlier, teosinte, the ancestor of corn, was among their food plants. These insects eventually became

dedicated eaters of corn and followed this crop northward as corn culture diffused into what is now the United States. This northward expansion of the western corn rootworm probably started over a thousand years ago, when the native people of what is now northern Mexico and the United States began to grow corn as a crop. Today, western corn rootworms have spread throughout the corn belt of North America.

The association between corn and northern corn rootworms came about much later and in a different way. These insects are natives of the tall grass prairie of the midwestern United States, and, according to F. M. Webster, probably had little or no contact with corn until some time after 1865, when European settlers began to grow this crop on the prairies of the midwest. Then northern corn rootworms made a massive switch to corn from their original food plants, perennial prairie grasses such as big blue stem. Terry Branson and James Krysan believe this happened in an area centering on central Illinois and eastern Missouri. Since then northern corn rootworms have spread east and west so that they now occur throughout the corn belt.

After entomologists came to understand the ecology and behavior of the two corn rootworms, especially the virtually total reliance of the larvae on corn as their food plant, it became quite apparent that the harm done by these insects could be avoided by rotating crops, by not planting corn in the same field for two or more successive years. (This tactic was first suggested in 1883 by Stephen Alfred Forbes, State Entomologist of Illinois, first head of the University of Illinois Department of Entomology, and one of the great early entomologists.) Since these insects have only one generation per year and the parent beetles die after they lay their eggs in late summer and autumn, the tiny larvae that hatch from the eggs the next spring will die of starvation unless the roots of a living corn plant are nearby. Thus the soil in a field that had been planted with some other crop the previous year, but is now planted with corn—known as first-year corn—will contain no rootworm eggs and the corn will not be infested.

In the nineteenth and early twentieth centuries, corn was grown in rotation with hay, clover, alfalfa, or small grains. But, beginning in the 1920s, when soybeans began to replace these other alternative crops, the rotation became more and more an alternation between corn and soybeans. By the early 1990s in Illinois, 27 percent of the corn crop was planted in fields that had previously been in corn, but almost 67 percent was grown in rotation with soybeans and another 6 percent in rotation with other crops.

Crop rotation prevented corn rootworm infestations for a hundred years and continues to do so in many areas of the country. But all organisms are constantly evolving, and both species of corn rootworms have evolved strains that can, each in its own way, get around crop rotation. The northern corn rootworm was the first to find a way to circumvent crop rotation. Instead of terminating diapause after the first winter and thus hatching when only soybeans or some other unacceptable crop is available, many of the eggs now remain in diapause longer and do not hatch until the spring after the second winter, when corn is once again available in a rotation of corn with soybeans or some other crop. Entomologists refer to this phenomenon as extended diapause. Only a few scattered signs of extended diapause were seen during the first half of the twentieth century, but by the 1980s and 1990s extended diapause had become widespread and had resulted in losses to first-year corn in North Dakota, South Dakota, Nebraska, Minnesota, Iowa, Illinois, and Michigan. Northern corn rootworms with an extended diapause are spreading and will eventually be prevalent throughout the corn belt.

In 1962, H. C. Chiang reported that only 2 of 676 northern corn rootworm eggs (slightly less than 0.3 percent) collected in Minnesota continued in diapause through a second winter. But that soon changed. In 1984, James Krysan and two colleagues reported that 40 percent of 329 eggs laid by northern corn rootworm beetles collected in nearby South Dakota had an extended diapause. In 1992 Eli Levine and two colleagues found that of 311 eggs laid by northern corn rootworm beetles collected at the same site in South Dakota, over 50 percent of those that hatched had an extended diapause: 20.6 percent hatched after two winters, 20.9 percent after three winters, and 9.6 percent after four winters. They found that in Illinois a similar proportion of the eggs of these beetles had an extended diapause: of 777 that hatched, 42.1 percent, 8.0 percent, and 0.3 percent, respectively, hatched only after two, three, or four winters.

Western corn rootworms found a completely different way to circumvent crop rotation but not until several years later. As Levine and his colleagues reported, the first indication that western corn rootworms

had found a way to get around crop rotation was seen in 1987 near Piper City in east-central Illinois. Ever since then this insect's ability to circumvent crop rotation has been spreading from this focal point—by 1995 to 23 counties in nearby Illinois and Indiana. By 1997, rotation-circumventing western corn rootworms, presumably carried by westerly winds, were rapidly spreading eastward and had been found as far away as southern Michigan and northwestern Ohio. Joe Spencer, an entomologist with the Illinois Natural History Survey, confidently predicts that it will not be long before they reach the corn-growing areas of Delaware and Maryland. Their spread in other directions, which is not aided by the prevailing westerly winds, has been much slower.

Rotation-circumventing western corn rootworms synchronize with rotated corn by laying many of their eggs in soybean fields rather than in corn fields. They seem to have no special preference for laying their eggs in soybeans. They lay them in association with several other plants, but most of their eggs will, of course, be laid in soybean fields, because soybeans are by far the most abundant plants other than corn in most midwestern farming areas. Western corn rootworm larvae do not survive on a diet of soybean roots and although adult western corn rootworms will eat soybean leaves, they do not survive if they have nothing else to eat. If corn and soybeans were not planted in alternate years in the same field, the western corn rootworm's new behavior of laying its eggs in soybean fields would be suicidal, because the larvae that hatch from these eggs would not coincide with corn, which is practically their only food. But in a corn-soybean rotation, larvae that hatch after the first winter from eggs laid in soybean fields will coincide with the corn that will be planted the next year.

It is truly remarkable that northern and western corn rootworms have evolved two totally different ways of circumventing crop rotation and thus coinciding with their food plants. For a hundred years farmers who rotated corn with other crops avoided the depredations of corn rootworms. But northern and western corn rootworms have now found, through natural selection, two ways—a different way for each species—of circumventing crop rotation and coinciding with corn—thus avoiding what would for them be a catastrophe. Some people might object to my use of the term *natural*, arguing that in these cases selection is artificial, because the crops are rotated by people rather than by "nature."

You may use whichever term pleases you. But the fact remains that selection is occurring, and we are now seeing evolution in action.

❁ ❁ ❁ Corn rootworm larvae occur in loose groups because egg-laying females gravitate to the same resource. A corn rootworm larva gains little or no advantage from being in a group with others. But cooperation between workers in a honey bee colony makes them the champion insect discoverers and exploiters of resources: especially pollen and nectar, almost their only foods; but also nesting sites; water to air condition their nests; and plant gums to seal cracks in the nest. Because the workers in a honey bee colony cooperate with each other, they can efficiently utilize the various sources of pollen and nectar in their territory, synchronizing their efforts to coincide with sources that are constantly shifting in time and space. The plants in a patch will yield pollen and nectar while they are blossoming but will be useless when the blossoms fade. Honey bee colonies are likely to exploit several kinds of pollen and nectar plants at the same time, but if a colony is to survive and prosper, it must constantly find new food sources to replace others that become unproductive.

Honey bees live in large, complex, and highly organized societies, probably the most advanced of the insect societies, and among animals other than insects, exceeded in complexity only by human society. But honey bee and human societies are fundamentally different. A human community with a population of 50,000 consists of people and family groups that are usually not related to each other. A honey bee colony may include 50,000 individuals, but it is a huge family unit, all of its members the children of one queen, the only member of the colony that is capable of reproducing, although in the absence of a queen, normally sterile workers can lay unfertilized eggs that will become drones (males). The colony includes a few hundred or a few thousand drones and nearly 50,000 workers, all of the latter sterile females. Although the workers almost totally forgo the opportunity to reproduce, they increase their inclusive fitness by benefiting relatives that have many genes in common with them.

A honey bee colony is, for two major reasons, exquisitely prepared to exploit with maximum efficiency the pollen and nectar sources in its ter-

ritory, which often encompasses as much as 43 square miles. First, the colony constantly sends out workers that scout for new sources of pollen and nectar. Second, when a successful scout returns to the nest, she uses the "dance language of the bees," more precisely the waggle dance, to communicate to her sister workers the exact location of the new resource. The language of the honey bees is one of the most astonishing marvels of nature. Until Karl von Frisch deciphered the meaning of the waggle dance early in the twentieth century, no one even dreamed that one insect could communicate complex information to another insect.

The waggle dance is performed on the vertical surface of one of the waxen combs that hangs in the dark interior of the nest. Well described by Thomas Seeley in *Honey Bee Ecology* and *The Wisdom of the Hive,* the dance is in the form of a figure eight. The scout begins by making a straight-line run, the crossbar of the eight. She then turns to the right and loops back to the beginning of the crossbar and makes another straight-line run along its path. Then she loops to the left and again returns to the beginning of the crossbar. She has now described a figure eight and repeats this pattern time after time, waggling her abdomen from side to side and emitting a high-pitched squeak as she moves along the crossbar. As she dances, other workers crowd in close, touch her with their antennae, and follow her through the movements of the dance.

The straight-line waggle run of the dance conveys two crucial pieces of information, the distance to the new resource and the direction in which it lies. The distance is conveyed by the waggle run's duration—the longer it takes, the more distant the flowers. The direction is expressed—quite accurately—as a deviation from a straight line from the nest to the sun. Just as we follow the convention that the top of a map represents north, honey bees follow the convention that the top of the comb symbolizes the direction of the sun. Thus when a worker conveys the direction of a food source that is directly opposite the direction of the sun, she directs the waggle run straight downward. If its direction is directly toward the sun, her waggle run goes straight up the comb. If the direction is 50 degrees to the left of a straight line from the hive to the sun, the waggle run is directed 50 degrees to the left of the vertical axis of the comb.

The workers that have followed the scout's dance then fly to the new patch of blossoms. If they are successful in their quest for pollen and nectar, they return to the hive to repeat the dance and recruit more workers. If the patch of blossoms is not productive, they do not repeat the dance. As long as the source continues to be productive, more and more workers will forage there and dance to recruit yet more workers. In this way, the foraging activity of the colony is focused on this patch and probably a few other productive patches of blossoming plants.

✿ ✿ ✿   The waggle dance of the honey bees is only one example, albeit the most sophisticated one, of how animals find the food they must have to live. Those that are predaceous must catch and subdue the animals they eat. Even herbivores must "subdue" the plants on which they feed in the sense of coping with toxins contained in the plants, bypassing protective spines or hairs, or even figuring out how to get at the nectar in a blossom. As you read on, you will find that sometimes groups may more effectively "subdue" food than can single individuals. A few predaceous insects and spiders increase their efficiency as hunters by ganging up on their prey, and in a few cases, insects may overcome the physical or even chemical defenses of plants through group efforts.

Subduing Food

*Jack pine sawfly larvae feeding on the needles
of their host plant*

⚙ ⚙ ⚙ Many insects and other arthropods are solitary, loners that do not associate with others of their kind except when they mate. There are among them a great many lone hunters, some of them miniature counterparts of predaceous vertebrates. A large and colorful ground beetle, sometimes known as the fiery hunter, climbs a tree to search for the caterpillars that are its natural prey. A crab spider that changes its color to match the white, petal-like sepals on the Canada anemone blossom on which it lurks remains motionless and almost invisible as it waits to grab a visiting bee or fly. After dark a wolf spider prowls the ground in an open field searching for an insect victim that it will pounce upon and subdue with its venomous bite. On a green plant, sometimes near a blossom, a large praying mantis waits motionless to ambush a passing insect—on rare occasions even a hummingbird—that it will snatch with its menacing, grasping front legs. In a sandy patch of soil— frequently one protected from rain by an overhanging branch or rock— an antlion larva, also known as a doodlebug, lies hidden in the dry sand at the bottom of the funnel-shaped pitfall trap that it dug. It waits to suck the juices from any unwary insect that tumbles down the steep slope of the pitfall, its descent often hastened by barrages of sand that the antlion flings at it with its shovel-shaped head.

There are tens of thousands of other insects, spiders, and related arthropods that are hunters, but, with the exception of some of the ants and spiders, almost none of them hunt in groups. The ants are exceptions because they live in closely knit societies, which are actually large families, and have evolved behaviors that enable them to live in harmony with their nestmates, all of which are their sisters, and to cooperate with them. Cooperative hunting very rarely occurs in other predaceous arthropods, despite the fact that the members of a group—even a small one—would under many circumstances be more successful than a solitary hunter. A group can usually find more prey and subdue larger and fiercer prey than can a lone individual.

There are vertebrate predators that hunt alone: the solitary mountain

lion that chases a deer in the Kaibab forest of Arizona, the crocodile that waylays an antelope coming to an African river to drink, the great blue heron that stands stock-still at the edge of a Wisconsin lake as it waits to seize a fish that comes too close. But many predaceous vertebrates improve their chance of making a kill by ganging up on their prey: the pack of wolves that overcomes a huge and powerful moose on Isle Royal, the closely knit pod of killer whales—orcas—that kills and devours a much larger whale, the pride of lions that cooperate to bring down a zebra, the small flock of white pelicans that swim abreast on a lake in British Columbia as they beat the water with their wings to herd along a school of small fish that they scoop up in the huge pouches on their bills.

Why do so few insects and spiders hunt in groups, while group hunting is more frequent among vertebrates, especially mammals? No one knows. But it may be because predaceous insects and spiders are far more likely than not to be cannibalistic, to turn on and eat members of their own species. Many are indiscriminate hunters that strike at or pounce on almost anything, including a fellow member of their own species, that moves and is large enough to make a worthwhile meal and small enough not to be frightening. A few plant-eating insects are cannibalistic, but most are not. There are numerous examples of cannibalism among predaceous insects and spiders, although not all of them are cannibals. The first meal of one of our North American praying mantises, for example, is likely to be one of its brothers or sisters that hatches from the same egg mass. When they grow larger, they will strike at and eat almost anything that moves—even other mantises. Green lacewings, also known as aphidlions, lay their eggs in groups on leaves or twigs. As Walter Balduf wrote, the eggs are widely spaced and each is at the top of a tall, thin stalk. This arrangement probably serves to keep the highly cannibalistic, newly hatched larvae away from each other. Males of predatory species are often eaten by their mates. Some males, such as mantises, are resigned to this fate, continuing to copulate even as their mates munch on them. But the males of most predatory insects try to avoid being eaten. Some male insects even bring their carnivorous mates nuptial gifts of dead insects to sate their appetites. And many male spiders use elaborate signals as they approach a female to inform her that they are suitors rather than a snack. Even so, they may be eaten if they don't make a hasty retreat after inseminating her.

Other than ants and some social wasps, there are very few predaceous insects that tolerate—let alone interact with—members of their own species except during mating. These few, a very few, give their young a modicum of parental care; they guard their eggs and may even watch over their newly hatched offspring for a short period of time. Among them are a few mantises and assassin bugs. As you will read below, egg guarding and watching over newly hatched young is much more common among spiders, despite the fact that they are all predators.

While parental care is rare among predaceous insects, it is relatively common among insects that are not carnivorous and thus not likely to be cannibalistic, such as scavengers and plant eaters. The situation among the rove beetles points up this difference. Of the several thousand known species of rove beetles, almost all are predators and none of the predaceous species are known to give parental care of any sort. But a few species of rove beetles feed on algae or dung, not prey. These species do give parental care, both guarding their eggs and caring for their young. In these nonpredaceous rove beetles, predatory behavior and cannibalism were not impediments to the evolution of parental care.

An important question remains unanswered. Why have so few predatory arthropods—compared to the predatory vertebrates—evolved ways to control their cannibalistic tendencies and thereby make it possible to reap the benefits of hunting as a group? Some arthropod predators might not benefit from cooperative hunting. But others might. After all, a group, with its several pairs of eyes and several "noses," may be better able to find prey than a lone hunter. The group would very likely be able to overcome larger prey, and its members might be better able to protect themselves against the attacks of other predators than would a lone individual. Biology, including sociobiology, has no answer to this question. But bear in mind that living in groups is not at all foreign to insects. Many plant-feeding and other nonpredatory insects are gregarious during at least some part of their life history. They do not have the problem of evolving a way to keep themselves from eating other members of their own species. But predaceous insects and spiders obviously

do have this problem. For some as yet undiscovered reason or reasons, it is more difficult for arthropods to relinquish through the evolutionary process the indiscriminate or even cannibalistic aspect of their prey-catching behavior.

⚙ ⚙ ⚙   Ants are descended from predatory wasps, and most ants have retained the predatory habit, although some have switched to other foods such as seeds or fungus grown in their nests. All ants are social and live in colonies. Generally speaking, a colony consists of one or more queens and anywhere from a few dozen to many thousands or even millions of workers. The workers, as is the case with social wasps and bees, are all females and all sisters or often half sisters, daughters of a single queen who may have been inseminated by several males. The ants are predisposed to group hunting because their highly organized societies cannot exist unless all members of a colony can recognize each other, refrain from eating each other, and cooperate to accomplish their many communal tasks.

In many species of predaceous ants, several workers cooperate to subdue a large insect and carry it back to the nest. In a few species, however, the workers are solitary hunters. The workers of an ant of southern Asia (*Harpegnathos saltator*), for example, are large enough and venomous enough to subdue and carry back to the nest by themselves insects as large as cockroaches, butterflies, cicadas, and grasshoppers. Some ants that are usually solitary foragers recruit helpers when they encounter an insect that is too large for them to handle. The solitary worker lays out a recruitment trail, marked with a pheromone from her poison gland, that other workers follow to where she waits with the too-large-to-handle insect.

There are many ants that hunt and retrieve prey cooperatively, but among the most remarkable of them are the swarming army ants of the forests of the New World tropics, especially *Eciton burchelli*, the most famous and the best understood of them. These ants are migratory, living in temporary camps or bivouacs located in more or less sheltered sites, sometimes under a fallen tree trunk or on the trunk or a main branch of a standing tree. The bivouacking ants build no shelter. The queen and the immature ants, eggs, larvae, and pupae, are surrounded and sheltered by a large, more or less ellipsoidal mass of many layers of workers, all with their legs and bodies linked tightly together. Such a bivouac may consist of anywhere from 150,000 to 700,000 workers and may be more than 3 feet in diameter. During the ants' nonmigratory phase, a bivouac usually remains in the same place for 2 or 3 weeks.

Each morning at daybreak many of the workers leave the bivouac to forage. The swarm moves out as a column that ultimately fans out to form a front that may be as much as 60 feet wide. The small and medium-sized workers (minimas and medias) race ahead along the chemical trail previously laid down by their sisters, extending it at its point. The column is flanked and defended by soldiers with large heads and very long sickle-shaped mandibles. The workers find and kill prey, mostly insects, and carry it back to the bivouac. As Bert Hölldobler and Edward O. Wilson put it in *The Ants,* workers often cooperate as teams to transport prey. The large medias, which act as porters, initiate the moving of large prey items and are joined by other workers of the same or smaller size. "The teams accomplish their task with greater energetic efficiency than if they cut the prey into small pieces and carried them individually."

As T. C. Schneirla wrote, "The huge sorties of *burchelli* . . . bring disaster to practically all animal life that lies in their path and fails to escape. Their normal bag includes tarantulas, scorpions, beetles, roaches, grasshoppers, the adults and brood of other ants; and many other forest insects; few evade the dragnet." Schneirla notes that snakes, lizards, and nestling birds may be killed by stinging or asphyxiation. "But lacking a cutting or shearing edge on their mandible, unlike their African relatives the 'driver ants,' these tropical American swarmers cannot tear down their occasional vertebrate victims." A raid by army ants is a conspicuous and dramatic event in the forest. The swarm is followed by noisy birds of several species, which do not eat the ants but appropriate for themselves some of the insects that are flushed by the ants. Fluttering over the scene are certain butterflies that suck up juices from the birds' droppings but are not eaten by the birds because they are toxic and warningly colored.

At dusk the ants return to the bivouac and resume their places in the mass of their sister ants. This behavior, repeatedly swarming during the day and returning to the same bivouac at night, is known as the

"statary," or nonmigratory, phase of the colony, and, as I have said, it persists for 2 or 3 weeks. The migratory phase of the colony follows the statary phase, also lasts for 2 or 3 weeks, and ends with a return to another statary phase.

The emigration from the statary bivouac begins at about dusk, when, as Hölldobler and Wilson wrote, "workers stop carrying food into the old bivouac and start carrying it, along with their own larvae, in an outward direction to some new bivouac site along [their] pheromone-impregnated trails." The queen doesn't make the journey until well after dark, by which time most of the larvae have been carried to the new bivouac site. As she runs along in the moving column, she is surrounded by and sometimes totally enveloped by a mob of soldiers ready to protect her with their lives. After her safe passage, the emigration gradually diminishes and is usually completed by midnight. During the migratory phase, the colony continues to hunt during daylight, but moves to a new bivouac site at about dusk at the end of each day.

These switches between phases are triggered not by the shortage of food but rather by the rhythmic cycle of reproduction that occurs within the colony. During the statary phase the colony includes a large number of developing pupae and the queen lays as many as 300,000 eggs. Toward the end of this nonmigratory phase, the eggs hatch more or less synchronously and a few days later new adult workers emerge from the pupae, also more or less synchronously. This triggers the beginning of the migratory phase, during which a fresh supply of prey to feed the ravenous larvae becomes available every day. But as soon as the larvae molt to the nonfeeding pupal stage, thus greatly decreasing the amount of food required, the colony reverts to the statary phase.

🔅 🔅 🔅   Since all spiders are predators and many are cannibalistic—at least during the later stages of life—you might expect them to be as antisocial as the predaceous insects. But this is not the case. For one thing, many spiders—even some species that ultimately become cannibalistic—guard the rather large silken sacs that they spin to protect their eggs. Some of these egg guarders stay with their newly hatched spiderlings for different lengths of time. The young spiders eventually disperse, but during this tolerant phase the mothers do not eat their young and the spiderlings do not eat each other. Some mothers go so far as to feed their tiny young, allowing them to share their prey or even regurgitating food for them. This noncannibalistic parental behavior may have predisposed spiders to become gregarious and even to capture their prey cooperatively. A mother and her family could, at least in theory, stay together indefinitely as a group. In other cases, this toleration of members of their own species, carried forward to adulthood, could make it possible for even unrelated individuals to join together and capture prey cooperatively. According to Ruth Buskirk, it is very likely that both of these evolutionary routes to group living have been followed.

Although the great majority of spiders are solitary, there are some gregarious ones—all of them web-spinning species. Some are gregarious in the simplest sense. They spin individual webs that are very close to each other and even touching. But each spider rests in its own silken shelter, tends its own web, and by itself subdues and eats the prey that are trapped in it. However, these spiders at least tolerate each other and do not attack their neighbors even when they are hungry because prey are scarce. There are several gregarious spiders that are also not communal subduers of prey, but associate more closely with each other than the species whose behavior I have just described. For example, as many as a thousand individuals of *Philoponella republicanus,* close relatives of the species described above, may occupy a communal web that may be over 9 feet wide. Together the spiders spin a supporting tangle of threads in a tree and construct a communal retreat in which they rest and to which they flee when threatened by an enemy, but each spider spins its own little prey-catching web on the communal support.

According to J. Wesley Burgess, only a handful of species—just four or five—build communal webs and trap and subdue their prey cooperatively. One of these species is *Mallos gregaris,* which is commonly found in Mexico. Many of these small spiders—they are seldom more than 0.2 inch long—cooperate to build a colonial web and shelter that surrounds the branches of a tree and may be more than a cubic yard in volume. One of these nests consists of an outer covering of sticky silk that surrounds a warren of tunnels and chambers in which the spiders rest. Their sticky web traps mostly flies, although other insects are also captured and eaten. Observations of a laboratory colony showed that when a fly is stuck in the web, several spiders converge on the frantically

buzzing insect. They subdue the fly with bites that inject venom and then peacefully feed on it together.

In Mexico, this species is known as *el mosquero,* the fly killer. According to Ruth Buskirk, since pre-Columbian times people in central Mexico have used the nests of this species much as other people use sticky flypaper. In the rainy season, when house flies are most numerous and are often oppressively abundant invaders of homes, people in the countryside bring cut branches with *Mallos* nests into their homes to reduce the population of pestiferous flies.

Edward O. Wilson wrote that the West African spider *Agelena consociata* is as close as any other arthropod to being truly social, as are termites, ants, honey bees, and some wasps. Like most of the social insects, these spiders cooperate to rear their young. The young spiders occur in groups of 50 or more and are obviously the offspring of more than one female because a single female lays a mass of no more than 30 eggs. They cluster in their silken shelter and wait for food to be brought to them by adults, which feed all spiderlings in a group and apparently do not distinguish between their own offspring and the others. Web building and prey capture are also communal activities. A nest may contain several hundred individuals. It is woven in the shrubbery along the edge of a rain forest, usually consists of several adjoining horizontal sheets of silk—sometimes as many as six or eight—and is often over 3 feet wide at its greatest breadth. An exceptionally large web about 32 feet long has been reported. Beneath the web are many silken galleries and a central retreat for the spiders.

These spiders are invariably friendly to their nestmates, frequently touching each other with their legs and pedipalps (antenna-like appendages) without any apparent hostility. They will also accept members of their own species that are transferred from other colonies, suggesting that natural colonies may not consist solely of related individuals. Flying insects are caught when they collide with vertical supporting threads and fall onto the horizontal web below. Although the web is not sticky, unlike the webs of most spiders, the insects become entangled in its many threads. Alerted by the impact of a fallen insect, the spiders, guided by the vibrations made by the struggling insect, move toward it. If the insect stops struggling, the spiders stop moving because the vibrations no longer betray the location of the prey. Small prey are immobilized and eaten by the closest spider; but after it has begun to feed it may be joined by other spiders. Large prey are killed by groups of spiders, which then all join in to enjoy their communal meal.

⊛ ⊛ ⊛ Most plant-feeding insects must "subdue" the plants on which they feed. This idea seems strange or even absurd to us; after all, we don't have to struggle with spinach or broccoli. Compared to insects, people and other plant-eating vertebrates are large, range over relatively wide areas, and eat many different kinds of plants, just passing up those that fight back with thorns or toxins. White-tailed deer, for example, wander over many acres and eat over 50 species of plants. In ecological terms, they occupy a broad niche. (An animal's niche includes the place where it lives, the food it eats, and all of the other resources it requires.) Largely because they are so small, most insects can and usually do occupy much more restricted niches than do vertebrates. Almost all insects are far smaller than the smallest invertebrates. A tiny white-footed mouse weighs less than an ounce, but weighs about 8,000 times as much as an insect of average size. Insects are small enough to live in, both figuratively and literally speaking, the "cracks and crevices" of the environment. There is, for example, a blood-sucking louse that spends its whole life in one nostril of a seal. Tiny insects known as leaf miners burrow between the upper and lower surfaces of a single leaf throughout their larval stage.

Some plant-feeding insects, among them armyworms, Mormon crickets, and grasshoppers, feed on many different kinds of plants. But most other plant-feeding insects, tens of thousands of them, have far narrower niches, and feed on only one or a few closely related species of plants that belong to the same family. They are host-specific. (The only host-specific mammal is the koala of Australia, which will feed only on some species of eucalyptus.) Certain caterpillars eat only leaves of the catalpa tree; cabbage aphids suck sap only from cabbages and other members of the mustard family; and the maggots of Hessian flies live only on a few grasses, especially wheat. An insect may spend its entire life—at least its entire growing stage—on just one individual plant. Just one needle on a pine tree is the whole universe of a tiny scale insect of a certain species. Judging from the few counts that have been made, prob-

ably the great majority of plant-feeding insects are host-specific. For example, Arthur Weis and May Berenbaum reported that 81 percent of the caterpillars of North American butterflies feed on plants from only one family.

✪ ✪ ✪ Plants have evolved ways to protect themselves against the insects that seek to attack them—ways that range from obvious physical defenses, hairs, thorns, or spines, to more subtle biochemical defenses. Plants, as a group, produce thousands of "secondary" biochemicals—called secondary because they are not essential to the life of the plant. Any one species produces only a few of these substances, and different plants of the same family generally produce the same or very similar secondary substances—thus enabling insects to recognize plants as belonging to the same family. Except for scents that attract pollinators, these secondary biochemicals, which are generally toxic or nasty tasting, protect the plant against insects and other attackers such as fungi, bacteria, and even mammals.

Plants are not protected against all insects. Almost every known plant is host to one or more insects that have evolved ways of coping with its defenses. There has been and there continues to be an "escalating evolutionary arms race" between insects and plants. If a plant comes up with a new defense, the insects that feed on it may go extinct unless they evolve ways to cope with the new defense, such as physiological mechanisms for breaking toxins down into harmless components. The plant may then respond by evolving yet another new defense. In this way insects become ever more closely bound to the plants with whose defenses they can cope. Many of them even come to be attracted by the tastes or odors of their host plant's secondary substances.

✪ ✪ ✪ Although predaceous insects very seldom occur in groups, many species of plant-feeding insects—generally but not always restrained by the fear of being eaten by another member of their species—occur in groups of from three or four to hundreds or even thousands or millions of individuals. Members of these groups may benefit in several ways, probably most often by escaping predators or having potential mates nearby. But only in a few instances has it been shown that a group can better cope with plant defenses than can an individual, though it is likely that many such instances remain to be discovered.

The group behavior of the tiny newly hatched larvae of the jack pine sawfly of the Great Lakes States and Canada alleviates the difficulty of biting through the tough outer wall of a pine needle to get at the softer tissues inside, as Arthur Ghent discovered. The eggs of this species, somewhat over a hundred per female, pass the winter in cavities in the needles of small clusters at the tip of a shoot. All of the larvae hatch at about the same time in May and soon try to feed on the needles. There is no doubt that it is difficult for the tiny larvae to bite into one of the tough-skinned needles. Of 312 newly hatched larvae, each confined alone in a test tube with pine needles, 80 percent died, almost always because they failed to penetrate the wall of a needle. But when 388 larvae were similarly confined in groups of 4 or more with pine needles, only about 50 percent of them failed to survive. The opportunity to form groups made the difference. When several larvae were together, a larva that managed to penetrate the tough surface of a needle—perhaps because it was stronger than most or happened to find a weak spot—was almost immediately joined by other larvae, some of which would certainly never have managed to penetrate a needle on their own. In nature, feeding groups of jack pine sawfly larvae usually consist of 7 or 8 individuals. These small feeding groups form within the larger group of 100 or more larvae that hatch from the eggs in a cluster of needles. With so many larvae present and crowded together, there will be some that can penetrate a needle and it will be easy for other larvae to find and join them.

❀ ❀ ❀   While jack pines are to some extent protected by the toughness of their needles, some other plants are physically protected by hairs or sharp spines. Some varieties of soybeans, for example, have smooth leaves, while others have densely hairy leaves. Smooth-leaved varieties are massively attacked by leafhoppers, tiny insects that suck sap from their leaves, and because of the injury done by the leafhoppers, seldom grow to be more than 8 inches tall. But leafhoppers cannot reach the surface of hairy leaves, although they readily feed

on them if they are shaved by an experimenter. Hairy-leaved varieties grow to be well over 36 inches tall. Needless to say, only hairy-leaved soybean varieties are grown commercially.

As Lawrence Gilbert recounted, the leaves of certain Costa Rican passion-fruit plants bear on their surfaces tiny, sharp, hooklike structures so closely spaced that the fleshy legs of a caterpillar that crawls onto one of these leaves will inevitably be snagged by many of these tiny hooks. The hopelessly trapped caterpillar eventually dies from a combination of blood loss and starvation. Gilbert postulated that this particular species of passion-fruit, one among many, has won the evolutionary arms race with the various species of caterpillars that specialize in feeding on plants of this family.

Although leafhoppers have yet to find a way to broach the hairy defense of soybeans, at least one caterpillar has found a way—one that requires cooperation between several members of a group—to avoid the sharp spines that protect the leaves of a certain member of the nightshade family, a plant that is closely related to the potato. This caterpillar belongs to a large family of butterflies that, so far as is known, feed only on plants of the nightshade family during their caterpillar stage. As Beverly Rathke and Robert Poole have pointed out, it is the only member of its family that feeds in groups; all others are solitary feeders. A group of these caterpillars, generally from four to six per leaf, spin a fine network of silk threads that covers the spines on the leaf and serves as a scaffolding on which they crawl over the tops of the spines to reach and feed on the unprotected edge of the leaf.

❁ ❁ ❁ Like the monarch, the orange and black milkweed bug (*Oncopeltus fasciatus*) is host-specific and feeds only on plants of the milkweed family. But while monarch caterpillars eat leaves, milkweed bugs pierce the plants' large pods to suck the contents of the seeds within them. Seeds are essential to them. If they feed only on vegetative parts of the plant, they grow very slowly and usually die. The developing seeds, each with a tuft of white floss on which it will ultimately float on the breeze, are protected by the tough, thick-walled pod, which splits open to release them only when they are fully mature. Almost all of the immature developing seeds are protected by the pod wall, most of which is so thick that the piercing beaks of the bugs cannot penetrate it to reach the seeds. But at the pod's "suture," the line along which it will split, the walls are much thinner, and the bugs can penetrate to the relatively few seeds in this area.

Milkweed bugs are gregarious, and as Carol Ralph wrote, they are more likely to survive when they feed in groups than when they feed alone. While the pod is closed they pierce its wall to feed on immature seeds, but when the pod begins to split open, they feed directly from the exposed mature seeds. But relatively few mature seeds are lost to the plant, because the pods soon open all the way and the seeds blow away. Jürgen Bongers and Wolfgang Eggerman have shown in the laboratory that milkweed bugs continue to be gregarious when they feed on mature seeds, and that three or four of them often pierce the same seed. According to Bongers and Eggerman, a sufficient quantity of saliva must be injected into a seed—especially a large dry one—to put its contents into a form that a bug can ingest. A lone milkweed bug cannot secrete enough saliva to accomplish this, but a group of bugs feeding together on a seed pool their saliva, injecting enough of it to render the contents of the seed ingestible and perhaps also digestible.

❁ ❁ ❁ Bark beetles (family Scolytidae) are the only insects known to me that depend upon group behavior to cope with the biochemical defenses of their host plant. These beetles are inconspicuous: plain brown and very small, rarely more than a quarter of an inch long. They are certainly among the most unprepossessing of the beetles. Since they spend most of their lives under the bark of a tree, they are seldom seen. Despite being elusive and unimpressive in appearance, they are among the ecologically most important insects in forest ecosystems.

When food-seeking adult beetles find a tree to their liking, they bore through the corky bark of the trunk to the interface between the bark and the wood. There they feed as they burrow along the interface, forming a tunnel partly in the wood and partly in the bark. This initial tunnel, the egg gallery, may be made by a polygamous male who will try to attract a harem of several females. In other species the egg gallery is prepared by a monogamous female and occupied by her and one male. In either case, the eggs are placed at more or less regular intervals along

the egg gallery. Each newly hatched larva begins its own tunnel at the bark-wood interface. At first it tunnels at a right angle to the egg gallery, but as it burrows and feeds, its tunnel widens and usually curves away from its right-angle path. The egg gallery and larval burrows are etched both into the wood and into the inner bark, as can be seen when a piece of bark is removed from the trunk. For this reason, the bark beetles are also known as engraver beetles. When the larva is fully grown, it burrows out into the corky layer of the bark to excavate a chamber in which it molts to the pupal stage. When it finally becomes an adult, it gnaws its way out of the bark to disperse and find a tree in which to start its own family.

The invasion of a tree by bark beetles is not as simple as my outline suggests. The tree fights back with formidable biochemical defenses, but the beetles sometimes manage to overcome these defenses through mass attacks in which they are aided in complex ways by mites and several species of fungi.

As adult and larval bark beetles burrow just beneath the bark, they destroy two tissues, each of which is essential to the life of the tree. One is the cambium, the all-important growth layer, which is only a single cell thick, but forms a complete cylinder that surrounds the trunk and extends out into the branches. The other essential tissue is the phloem, a complex of conductive tubes formed by the cambium, which lies between the cambium and the corky bark. Like the cambium, the phloem forms a complete cylinder around the trunk and branches. The phloem tubes carry nutrients, mainly sugar formed by photosynthesis in the leaves, down to the root system. Trees die if their trunks are girdled—if even a narrow but complete ring of cambium and phloem is removed from the trunk. Bark beetles probably seldom girdle trees, but they can destroy enough cambium and phloem to weaken a tree.

Bark beetles are ubiquitous components of forest ecosystems, and perform ecological functions that are essential to maintaining the diversity and long-term health of forests. But sometimes, especially when they become unusually numerous, they conflict with the short-run economic interests of people. On balance and in the long run, bark beetles surely do the forests more good than harm and enhance long-term productivity, but that is a view not shared by people who think only in the short term and see a forest as nothing more than an expendable producer of lumber. On occasion, bark beetles have become exceedingly numerous and have killed thousands or even hundreds of thousands of trees. The forest eventually recovers—just as the forests that burned in Yellowstone Park in 1989 are now well on the road to recovery. However, there is no denying that the short-run loss of harvestable trees can be great. F. P. Keen, a well-known forest entomologist of the early twentieth century, described a massive outbreak of a bark beetle in the west.

> The Engelmann spruce beetle in 1942 began to increase its numbers in wind-thrown spruce in the high mountains of western Colorado. From the windfalls, new beetle progeny emerged to attack living trees. With each generation, increasing hordes of beetles developed and attacked more and more spruce, until more than 4 billion feet has [*sic*] been killed.

During the flight period of 1949, the air was so full of beetles that the ones that happened to fall into a small lake in the infested area and then washed ashore (although only a minor fraction of the total flight) were so numerous as to form a drift of dead beetles 1 foot deep, 6 feet wide, and 2 miles long. Those flying southeast over 18 miles of open country settled on a plateau of previously uninfested forest and killed 400,000 trees in one mass attack.

Although forests recover from such destruction, in the short run some of the bark beetles are so economically important that great efforts have been made to understand their relationships with the trees that they attack. A great deal has been learned, and much of that has been summarized in a volume edited by T. D. Showalter and G. M. Filip. But much more remains to be discovered.

Bark beetles can be divided into two groups. There are nonaggressive species that attack sick or weakened trees. They have no need to form aggregations to overcome the relatively feeble defenses of these helpless trees. These beetles are attracted to declining trees by their odor, especially the odor of ethyl alcohol, a product of fermentation. Then there are the aggressive species, which are not attracted from a distance by odors but land on the trunks of trees at random and decide whether or not to attack the tree only after they have landed, apparently by means of chemical cues that cannot be perceived from a distance. They must determine if the tree is a species that is on their host-specific menu and if

it is weak enough to be easily attacked. These aggressive species, the subject of the rest of this discussion, often attack weakened trees, but when their populations are high, they can overcome the defenses of healthy trees by means of mass attacks in which thousands of individuals attack a single tree.

The trees are not helpless and can usually defeat the weak attack of a small number of bark beetles. The immediate response of the tree is to flood the attacking beetles with resin that may drown them in their burrows or render them helpless by entrapping them. The resin contains chemical compounds, various types of phenols and monoterpenes, that are toxic, repellent, or both. The tree also surrounds the site of the attack with a layer of dead wood that does not nourish the beetles and decreases the concentration of nourishing sugars at the site of attack. Pines attacked by the southern pine beetle of the southeastern United States, an aggressive species, also decrease their production of odorous alpha-pinene, which attracts these beetles and also enhances the effectiveness of their aggregation pheromone, which is known as frontalin.

The beetles respond to the tree's defense with a two-pronged attack. First, they produce aggregation pheromones whose odor can attract more beetles of their own species from a distance. The new recruits help to weaken the tree by destroying much of its conductive tissue, the phloem. Second, the bark beetles inoculate the tree with the spores of several fungi that help to weaken the tree or may even kill it.

Both males and females produce the aggregation pheromones. The beetles of at least some species synthesize their pheromones from breakdown products of some of the toxic compounds produced by the tree. In many cases, volatile compounds released by the tree itself synergize the action of these beetle-produced pheromones, just as alpha-pinene increases the effectiveness of the frontalin of southern pine beetles.

If the bark beetles do not attack in large enough numbers, the tree will defeat them. But if the beetles mount a massive attack, they and the fungi together can subdue even a healthy tree. The beetles then breed in the weakened and dying tree. They do not remain in dead trees. Other insects, including many species of beetles, live in dead wood and help the process of decay that will return the nutrients in the dead tree back to the soil.

The bodies of many adult bark beetles bear "pockets"—which in some species are actually nothing more than depressions—whose function is to carry spores of the fungi with which the beetles inoculate the trees they attack. As these fungi grow, most of them weaken their host tree but seldom kill it. But others, among them the deadly blue stain fungus, actually do kill trees.

Southern pine beetles have spore pockets on the front part of the thorax that contain the spores of more than one fungus, but do not contain spores of the blue stain fungus. Amazingly enough, the spores of this fungus are usually carried on the bodies of tiny mites that hitch rides on the bodies of southern pine beetles. Newly matured beetles that disperse from fungus-infested trees generally carry with them mites and their associated blue stain fungus spores. When one of these beetles burrows into a tree, the hitchhiking mites leave its body to move into its tunnel, carrying the blue stain fungus spores on their bodies. Once the blue stain fungus begins to grow, the mites spread it by transporting it throughout the beetles' galleries.

❂ ❂ ❂   Locusts, the subject of the next chapter, have evolved physiological mechanisms for overcoming the biochemical defenses of many different kinds of plants. When plants are abundant, they feed on only certain favored species, but when food is scarce they are capable of eating and digesting almost any species of plant, all but the most toxic ones. Locusts sometimes become so abundant that they devour all or almost all of the plants around them. Then huge swarms of them migrate in search of food. They are without doubt the most obvious and awesome example of insects "subduing" food.

Legions of Locusts

Migratory locusts; the female at the bottom embeds her abdomen in the soil as she lays a clutch of eggs

✵ ✵ ✵ Frozen in the ice of a glacier high in the Beartooth Mountains of Montana, as Jeffrey Lockwood and his coauthors reported, are the remains of thousands of Rocky Mountain locusts. They were members of a migrating swarm that was probably blown up onto the glacier between 1130 C.E. and 1340 C.E., as determined by radiocarbon dating. If they had escaped their icy fate, this swarm might have made its way eastward, as did many similar swarms in the nineteenth century, swarms that sometimes included billions of voracious plant-eating locusts. If it had, it might have devastated the crops of a native American village in the bottomlands of a midwestern river such as the Missouri. For by 700 C.E., many of the people in this area had become almost completely dependent upon agricultural crops, especially corn. These farmers would have considered this invasion of locusts a plague, an awful affliction. But a thousand years earlier, before agriculture had become an essential part of their culture, the ancestors of these people might have thought of such an invasion as more of a boon, a welcome opportunity to add to their food supply an abundant harvest of protein-rich locusts.

What is a locust? It is simply a grasshopper—a grasshopper with some unusual habits, but still a grasshopper. Most grasshoppers are stay-at-homes that do little or no migrating and are usually not particularly gregarious. But locusts, at least under certain conditions, are so gregarious that they stumble over each other and often migrate for hundreds of miles in immense flying swarms that may consist of tens of millions or even billions of individuals.

As Curtis Sabrosky wrote, in North America immense swarms of Rocky Mountain locusts swept over the Great Plains in the 1870s, the same species that was frozen in the ice of Grasshopper Glacier in Montana:

> The locusts are said to have left fields as barren as if they had been burned over. Only holes in the ground showed where plants had

been. Trees were stripped of their leaves and green bark. One observer in Nebraska recorded that one of the invading swarms of locusts averaged a half mile in height and was 100 miles wide and 300 miles long. In places the column, seen through field glasses and measured by surveying instruments, was nearly a mile high. With an estimate of 27 locusts per cubic yard, he figured nearly 28 million per cubic mile. He said the swarm was as thick as that for at least 6 hours and moved at least 5 miles an hour. He calculated that more than 124 billion locusts were on the move in that one migration.

When I first read this account I wondered how the observer in Nebraska had measured the length and width of this great swarm, an astonishing 300 by 100 miles. Measuring the length was easy. It was necessary only to multiply the swarm's forward speed by the time it took it to pass a given point. (Note that the 6 hours mentioned is only the time during which the *thickest part* of the swarmed passed.) How the width of the swarm could have been measured had me stumped for a while. But then I remembered that the settled parts of North America were already criss-crossed by telegraph lines in the 1870s. The width of the swarm could easily have been determined by sending a series of telegraphic queries across the country.

Given the instrumentation available in the 1870s, I can't help wondering about the accuracy of this estimate. The parameter that is the most difficult to measure is, of course, the density of the locusts in the swarm, the number per cubic yard. In the light of the densities of desert locust swarms obtained in Africa by sophisticated photographic techniques, about 12 per cubic yard, the figure of 27 Rocky Mountain locusts per cubic yard seems high, although they are much smaller than desert locusts and more could crowd into the same space. But even if we halve the density of the Nebraska swarm, we are still left with an impressive count of over 60 billion locusts.

❀ ❀ ❀ In the Rocky Mountains of the United States there are, as Lockwood and his colleagues reported, at least 12 mountain glaciers that have grasshoppers embedded in them. But as of 1990, only one of them, Grasshopper Glacier, had been examined by a biologist. The grasshoppers in the other 11 glaciers are probably also Rocky Mountain locusts, but we have no inkling of when they were frozen in the ice. Some might be much older than the ones frozen in Grasshopper Glacier about 800 years ago. Swarms of Rocky Mountain locusts probably flew in North America for many millennia before Europeans—and perhaps even native Americans—arrived, most likely at least since the last Ice Age ended about 14,000 years ago.

As Charles Valentine Riley noted in 1877 in *The Locust Plague in the United States,* the written history of the Rocky Mountain locusts began in 1819, when "vast hordes of [them] appeared in Minnesota, eating everything in their course; in some cases the ground being covered with them to a depth of three or four inches." These insects became ever more troublesome to people as more and more areas of the plains and prairie states were opened to farming by early settlers. Locusts that once descended upon wild prairie plants now devoured crops and threatened vast areas with famine. The worst of these invasions occurred in 1874, when swarms of Rocky Mountain locusts moved over a huge area about 600 miles wide that was bounded on the west by the Rocky Mountains, extended east to western Manitoba, Minnesota, Iowa, and Missouri, and extended south to central Texas. Riley reported that the monetary loss caused by locusts in 1874 was about $50 million. This is, of course, a far larger amount in 1998 dollars, probably several billion.

Rocky Mountain locusts were recorded as appearing somewhere in central North America every few years until 1879, the last year in which swarms—and then only minor ones—were seen. The species is now presumed to be extinct by most entomologists—for reasons that have yet to be discovered—but it has been proposed that the destruction of the prairie by the plow was involved. It should also be noted that the demise of the bison herds, which kept much of the prairie grass cropped short, occurred at about the same time that these locusts disappeared. Perhaps the locusts preferred to lay their eggs where the grass had been cropped.

Rocky Mountain locusts, as is true of all other locusts, were doubly destructive: as winged, migrating adults that fed heavily to sustain their long flights and to develop eggs; and as wingless nymphs that fed even more voraciously as they increased their weight several hundred times

as they grew from tiny, newly hatched hoppers to inch-long winged adults.

Everywhere they appeared, from Manitoba to Texas, the swarms, which invaded from the northwest, laid eggs by the millions in favorable places. The invading locusts appeared in August in Missouri, but earlier to the north; June in South Dakota, and later to the south, not until the middle of October in the vicinity of Dallas, Texas. Each female produced about 300 eggs, divided among 10 or so separate pods of about 30 eggs each that were buried an inch or more below the surface of the soil.

All female grasshoppers, including the various species of locusts, lay their eggs in essentially the same way. At the tip of the female's abdomen is an ovipositor, an egg-laying organ, that has short, broad blades with which she digs into the soil. As the ovipositor burrows down, to a depth of as much as 6 inches in one species, the abdomen is stretched ever longer as segments that are usually telescoped into each other pull apart but remain joined together by the broad, flexible membranes that had been folded between and hidden by adjacent segments but are now stretched to their limit. As she lays her eggs, the female spreads over them a frothy liquid that soon hardens to cover and bind them together as a pod. The eggs of the Rocky Mountain locust, like those of most North American grasshoppers, were in diapause and did not hatch until the following spring.

By early May of 1875, the eggs laid by the locusts that had invaded Missouri the previous summer had hatched, but the nymphs had not as yet done much damage, because they were still too tiny to eat much and had not yet spread away from their hatching grounds. From the beginning, the nymphs were gregarious, often congregating, as Riley said, "in immense numbers in warm and sunny places. They often thus blacken the sides of houses or the sides of hills." As the month of May progressed, the nymphs grew, exhausted the nearby food supply, and then spread widely over the surrounding area and became alarmingly destructive. Riley described the havoc they wreaked: "By the end of the month [May] the non-timbered portions of the country most affected were as bare as in winter. Here and there patches of *Amarantus Blitum* and a few jagged stalks of Milkweed (*Asclepias*) served to relieve the monotony. An occasional oat field or low piece of prairie would also remain green; but with these exceptions one might travel for days by buggy and find everything eaten off, even to the underbrush in the woods. The suffering was great and the people were well-nigh disheartened."

Some people thought that the swarms of flightless nymphs were led by "kings" or "queens" that guided them in their peregrinations. This erroneous idea was based on the observation that a few large, winged, adult grasshoppers were occasionally seen with the crowds of nymphs. These adult grasshoppers were not royal leaders, not even Rocky Mountain locusts; as Riley explained, they were members of other species that diapause as adults and thus happened to be present as full-grown individuals in early spring when the locusts hatched from their eggs.

By early June, Riley reported, the locusts had reached the winged adult stage and had begun to leave the invaded area in swarms, flying toward the north and northwest, the direction from which their parents had come during the invasion of 1874. He wrote that this was a return migration to their native area near the Rocky Mountains, supporting his hypothesis by several observations that the locusts persistently adhered to their northwesterly course. They depended largely upon the wind for their progress, and Riley had heard reports that when it aided them by blowing from the south or southeast, they flew along with it, but when they encountered a headwind from the north or northwest that impeded their progress, they settled to the ground and waited for the wind direction to change.

⊕ ⊕ ⊕   Swarm-forming locusts of many different kinds are found almost everywhere on earth—on all of the continents except Antarctica. Although Rocky Mountain locusts are long gone from North America, other grasshoppers of this continent sometimes form migrating swarms, but far smaller and less destructive than those of the Rocky Mountain locust. The American migratory grasshopper, not to be confused with the migratory locust of the Old World, can be very damaging and occasionally makes mass flights to new feeding grounds that may cover as much as 575 miles during a season. At least seven species of locusts plague Africa, one or more of them occurring almost everywhere

from the southern tip of the continent to Morocco and other countries along its northern limit on the Mediterranean coast. One of the most infamous of them, the migratory locust (*Locusta migratoria*), occurs throughout most of Africa and also from southern Europe east to China, Japan, the Philippines, and Australia. Even as I write, huge swarms of migratory locusts are devastating crops all over the island of Madagascar, the Malagasy Republic. The equally infamous desert locust (*Schistocerca gregaria*) has a more limited range but swarms over northern Africa and east to Arabia, Pakistan, and India, and sometimes invades Europe. A close relative of the desert locust, a different species of *Schistocerca*, plagues the grasslands of the southern part of South America. Australia not only suffers from the devastation wrought by its own native race of the migratory locust, but also undergoes the predations of another species, the Australian plague locust, which is unique to that continent.

The locusts of Africa and the Mideast are at least as destructive as was the Rocky Mountain locust in North America. The biblical accounts of the devastation that they cause, quoted in the next chapter, are not exaggerated. Throughout recorded history, from the days of the ancient Greeks to the present, people have written accounts that describe equally horrendous locust scourges. Although such descriptions are common, few quantitative measures of the destruction wrought by locusts have been made and recorded. But R. F. Chapman, a British entomologist, assessed the amount of damage done by locusts to pasture lands in tropical Africa. He calculated that while the cattle on such a pasture—only about six head per square mile—ate about 150 pounds of vegetation per day, a swarm of adult locusts on the same area would eat about 127,000 pounds per day, over 800 times as much as the cattle. This leaves little or nothing for the cattle to eat, and it would very likely take years for the pasture to recover from the devastation caused by the swarm.

The quantity of vegetation destroyed by a swarm depends upon its size and the voraciousness of its members. Various estimates of the sizes of swarms have been published, ranging from absurdly inflated "guesstimates" to reasonable estimates that have a basis in science. One of the more absurd estimates, unfortunately sometimes repeated in the popular literature, was published in *Nature* in 1889 and is quoted in its entirety below:

Locusts in the Red Sea

A great flight of locusts passed over the s.s. *Golconda* on November 15, 1889, when she was off the Great Hanish Islands in the Red Sea, in lat. 13°56N., and long. 42°30E.

The particulars of the flight may be worthy of record.

It was first seen crossing the sun's disk at about 11 A.M. as a dense white flocculent mass, travelling towards the north-east at about the rate of twelve miles an hour. It was observed at noon by the officer on watch as passing the sun in the same state of density with equal speed, and so continued till after 2 P.M.

The flight took place at so high an altitude that it was only visible when the locusts were between the eye of the observer and the sun; but the flight must have continued a long time after 2 P.M., as numerous stragglers fell on board the ship as late as 6 P.M.

The course of flight was across the bow of the ship, which at the time was directed about 17° west of north, and the flight was evidently directed from the African to the Arabian shore of the Red Sea.

The steamship was travelling at the rate of thirteen miles an hour, and, supposing the host of insects to have taken only four hours in passing, it must have been about 2000 square miles in extent.

Some of us on board amused ourselves with the calculation that, if the length and breadth of the swarm were forty-eight miles, its thickness half a mile, its density 144 locusts to a cubic foot, and the weight of each locust 1/16 of an ounce, then it would have covered an area of 2304 square miles, the number of insects would have been 24,420 billions; the weight of the mass 42,580 millions of tons; and our good ship of 6000 tons burden would have had to make 7,000,000 voyages to carry this great host of locusts, even if packed together 111 times more closely than they were flying.

Mr. J. Wilson, the chief officer of the *Golconda*, permits me to say

that he quite agrees with me in the statement of the facts given above. He also states that on the following morning another flight was seen going in the same north-easterly direction from 4:15 A.M. to 5 A.M. It was apparently a stronger brood and more closely packed, and appeared like a heavy black cloud on the horizon.

The locusts were of a red colour, were about 2 1/2 inches long, and 1/16 of an ounce in weight.

G. T. Carruthers

Carruthers reported that this swarm of desert locusts consisted of well over 24 *trillion* individuals that weighed in the aggregate 42.5 *billion* tons. Needless to say, these are preposterously inflated estimates. There are two readily apparent sources of error. According to Sir Boris Uvarov, in his time the undisputed world authority on locusts, a recalculation of the crude data collected by Carruthers and his shipmates suggests that the area covered by the swarm was only about a fifth of the original estimate. What's more, the estimate of 144 locusts per cubic foot is absurdly high. The density of the swarm may actually have been less than one locust per cubic foot. There is no way that 144 flying desert locusts could crowd into an air space of 1 cubic foot! They are over 2 inches long and have a wingspan of about 5 inches. But Carruthers' data are still very impressive even after they have been tamed by reestimation. The swarm that he observed probably consisted of about 40 billion locusts that weighed about 125,000 tons in the aggregate, figures that are more or less in line with more recent, scientifically based estimates. According to a more accurate estimate based on photographic techniques and published in 1958, several swarms that convened in one area of Africa consisted of about 50 billion desert locusts that all together weighed somewhat more than 115,000 tons. That is a still lot of insects!

So numerous can locusts be that the beating wings of an approaching swarm can be heard from a distance. In *Thalaba the Destroyer,* the English poet Robert Southey vividly described this sound:

> Onward they came, a dark continuous cloud
> Of congregated myriads numberless,
> The rushing of whose wings was as the sound
> Of some broad river, headlong in its course

Plunged from a mountain summit, or the roar
Of a wild ocean in the autumnal storm,
Shattering its billows on a shore of rocks,
Onward they came, the winds impell'd them on . . .

⚙ ⚙ ⚙ How much locusts eat depends in part upon their size. Grasshoppers tend to be fairly large, although they come in a broad range of sizes. The smallest one known, an African species that does not qualify as a locust, is only about a quarter of an inch long. The largest, an Amazonian species that is also not a locust, is about 8 inches long and has a wing span of about 9 inches. Fortunately, none of the locusts approaches this monster in size.

The length of a grasshopper is not a completely reliable measure of its true size, because the abdomen can vary significantly in length depending upon the degree to which its segments are extended or contracted. Weight can also be somewhat misleading; a fully fed individual weighs much more than one with an empty gut. Although weights can only be approximated, they are, nevertheless, useful measures. An individual's wingspan, however, is always the same. The largest of the locusts, judging by wingspan, is the Bombay locust of southwestern Asia, which has a wingspan of almost 5.5 inches. The desert locust isn't far behind with a wingspan of only slightly less than 5 inches. The migratory locust, with a wingspan of about 4 inches, is significantly smaller. The Rocky Mountain locust was the smallest of all, with a wing span of only about 1.7 inches.

According to what has been published, adult locusts, generally speaking, eat about their own body weight in vegetation every day. The average weight of a desert locust is about 2.75 grams, a little more than 0.1 ounce. The average weight of a migratory locust is less than half that, about 0.04 ounce. It follows that an average migratory locust probably eats less than half of what a desert locust eats. Little is made of this difference in size in discussions of the damage done by these two locusts. Both species occur in such astronomical numbers that swarms of either one often eat virtually all of every suitable plant they come upon. Thus the difference in their sizes is of little significance compared to their abundance.

The two species do, however, differ in their food preferences, although there is a big overlap and both will eat *almost* anything under starvation conditions. Migratory locusts prefer grasses and sedges—wild grasses, grains, corn (maize), sorghum, and even sugar cane—but will also consume some broad-leaved plants, including bananas and peanuts. The desert locust is much more catholic in its preferences, happily feeding on grasses of all sorts and a great variety of broad-leaved plants, including among others the leaves of citrus.

There are no insects—not even the voracious and relatively nondiscriminating locusts—that will eat any and all kinds of plants. Many plants are toxic to insects, an evolutionary gambit that protects them against plant-feeding insects. Some insects, said to be polyphagous, have evolved ways of getting around the chemical defenses of many different kinds of plants. But there are toxic plants that not even the broadly polyphagous desert locust will feed on. Among them is the neem tree, whose insecticidal properties have long been known in India. D. R. Bhatia and H. L. Sika, Indian entomologists, published a striking photograph of a green and intact neem tree standing next to a nontoxic tree that was completely defoliated—every leaf is gone—by desert locusts.

⚙ ⚙ ⚙ Locust swarms come and go. A species may form swarms every year for several years in succession and then seemingly disappear for years or even decades. There were, for example, huge swarms of desert locusts almost every year from 1940 to the mid 1960s, but for the next 20 years they were absent except for a few minor outbreaks. As M. P. Pener explained, this long locust-free period deceived some people into thinking that it was no longer necessary to know your enemy, to support basic research on locusts, because they were pests of the past and no longer a threat. They were wrong! Catastrophically damaging swarms reappeared in the 1980s.

But where are the locusts in the years when there are no swarms? This question was not satisfactorily answered until 1921, when Boris Uvarov published a taxonomic treatise on the genus *Locusta*, the genus to which the migratory locust belongs. In that publication he proposed that what had until then been considered to be two different species of *Locusta* are, in fact, the same species and that this species occurs in two phases which are quite different in appearance and behavior: a solitary phase and a gregarious phase. The gregarious, swarm-forming phase is *the* migratory locust that has been known for millennia. The other, the solitary phase, is not at all gregarious, and occurs only as sparse populations of widely scattered individuals; little was known about it at the time Uvarov published his treatise. Uvarov's "phase theory," which applies to all locusts, is important. It finally revealed the true nature of locusts and stimulated a great deal of new research that ultimately made it possible to pinpoint the sources of locust swarms, to predict outbreaks, and to institute control measures that are not always successful but are certainly better than total helplessness in the face of catastrophic outbreaks.

Locusts do, of course, have genes that allow them to assume the anatomical and behavioral characteristics of the *gregaria* phase, the *solitaria* phase, or intermediates between the two. But—and this is a big but—which phase a locust assumes is not genetically preordained. It is somewhat affected by the conditions that the locust's parents experienced, but it is largely determined by the "social" environment of the growing locust, specifically by how crowded it is by other locusts. This has been clearly demonstrated by many experiments. Locusts reared in isolation always assume the characteristics of the *solitaria* phase, and those raised in crowds assume some or all of the characteristics of the *gregaria* phase. Even more convincing is the fact that both phases can be produced from a single pod of eggs, all of which are the offspring of one mother. If half of the hoppers that hatch from an egg mass are raised isolated from other locusts, they become *solitarias*. If the other half are raised under crowded conditions, they become *gregarias*.

In nature, locusts enter the gregarious phase when, in a given area, the growing nymphs become so numerous that they cannot avoid contact with each other. This can happen over a period of years during which rainfall and other climatic conditions favor survival and reproduction by the locusts. Eventually the population grows so large that it begins to outstrip the food supply. As food becomes scarcer, once solitary locusts enter the gregarious phase as they inevitably become crowded together with other solitary individuals while gravitating to the few remaining patches of food. Once the wingless nymphs enter the

gregarious phase, they become attracted to each other and ultimately form huge "armies" that march overland as they denude an ever increasing area of its plants. They stay together when they become winged adults, forming the huge flying swarms that wreak so much havoc. If a swarm ends up in a barren desert or falls into the sea, the locusts will die. But if they land in a favorable place with abundant vegetation, they will lay eggs—countless millions of them that may give rise to another devastating population of locusts.

Locust populations of the *gregaria* phase revert to the *solitaria* phase when they are no longer crowded. This happens when the population is sufficiently thinned by the various mortality factors to which it is subject, such as the weather, a shortage of food, or parasites and predators. Egg pods will be few and far between, and the nymphs that hatch from them will not be crowded enough to be induced to enter the *gregaria* phase. Entomologists used to think that once a nymph has entered the gregarious phase it cannot revert to the solitary phase, but Sylvia Gillett of the University of York in England showed that this is not so. In the laboratory, the tendency for gregarious desert locust nymphs to come together once again to form groups declined rapidly after they had been isolated from each other for more than 24 hours. Locusts probably behave the same way in nature—a new and important consideration in forcasting the formation of swarms.

But what is it about being crowded that causes a developing locust to enter the gregarious phase? It can become aware of the individuals around it by seeing, hearing, feeling, smelling, or tasting them. Is one of these senses more important than the others, or do all of them work in concert? It turns out that the sense of touch is by far the most important. Developing locusts, or hoppers, that were reared in isolation and then constantly touched by a mechanical device for 7 hours later proved to be as gregarious, when placed with other locusts, as a control group of individuals that had been kept for 7 hours—either in the light or in the dark—in a crowd of other locusts. Isolated locusts that received no stimuli were far less gregarious. The mechanical device, designed by Peggy Ellis of the Anti-Locust Research Centre in London, was ingenious: it consisted of several wires that were suspended from a disk in a cylindrical container housing the solitary locust and that constantly stroked it as the disk was twirled by a wind vane driven by a gentle breeze from a fan.

The isolated hoppers stroked by the wires received no stimuli from other locusts. They could not see, hear, taste, or smell other hoppers. Touch alone caused them to become gregarious. Other experiments done by Ellis made the same point. Lone locust hoppers raised in crowds of other species of locusts, grasshoppers, or even woodlice (small land-dwelling crustaceans) tended to be more gregarious than hoppers raised in isolation. Although more recent work, such as that of M. P. Pener, indicates that pheromones may help things along, it remains true that touch alone induces gregariousness, and that pheromones are probably only of secondary importance. Experiments in which solitary locust hoppers did not become gregarious although they could see crowds of other hoppers indicate that the sight of other hoppers is not enough to cause a hopper to enter the gregarious phase. But other experiments show that visual contact is important in another way, in keeping locusts together once they have been induced to enter the gregarious phase.

In all locusts, the two phases differ in appearance and behavior. The magnitude of these differences varies between species, and some individuals are, to varying degrees, intermediates between phases. Typical, undiluted members of a phase are the two extremes of a continuum between *solitaria* and *gregaria*.

Generally speaking, *solitaria* and *gregaria* migratory locusts are strikingly different in color pattern. *Solitarias* are green and blend in with the plants on which they feed, a camouflage that probably gives them some protection from predators. *Gregarias*, by contrast, are brightly colored and conspicuous, usually mostly black with yellow or orange markings. Conspicuously colored insects are usually toxic and distasteful to birds and other vertebrate predators, but most predators relish locusts and gorge on them. R. F. Chapman hypothesized that since the conspicuous markings of *gregarias* are not warnings of inedibility, they may "facilitate visual interaction between the individuals in a band and so help to maintain gregarious behaviour and band cohesion."

There may also be differences in size and sometimes readily noticeable anatomical differences between the two phases. For instance, soli-

tary migratory locusts have a conspicuous crest on the back just behind the head, but this crest is absent from gregarious individuals. Other anatomical differences tend to be more subtle: slight differences in shape, the length of the wings, the number of antennal segments, and differences in proportions, such as the ratio of the main segment of the hind leg to the width of the head.

Differences in physiology and behavior are usually more important and often more noticeable than anatomical differences. Crowded immature females of the migratory locust have more fat than isolated ones, but this difference disappears by the time they have become adults and laid their first pod of eggs. Thus it seems that, as is to be expected, the more active gregarious individuals burn up more fat than the less active solitary ones. When they hatch from the egg, the offspring of gregarious females are heavier than the newly hatched nymphs of solitary females. As a result, young *gregaria* nymphs can survive longer without food than *solitaria* nymphs, a decided advantage when a band of nymphs has denuded an area of food. The most apparent behavioral difference is the nonsocial behavior of *solitaria*, compared to the extreme gregariousness of the other phase. Both as nymphs and adults, *gregarias* are obviously attracted to each other and stay together as more or less tightly packed groups.

❀ ❀ ❀　Locusts of all species may be attacked in all of their life stages by a variety of predators and parasites, among them cannibalistic individuals of their own species, other insects, mammals, small birds, storks that stand 3 feet tall, and even microscopic fungi. Relatively small populations of gregarious locusts are sometimes totally wiped out by birds and other predators. C. Ashall and Peggy Ellis mention a small population of locust nymphs that was destroyed in a week by flocks of starlings, weaver birds, and a few storks. Although the nymphs and adults in large swarms are ravenously devoured by birds and other predators, the size of these swarms is not significantly diminished because there are far more locusts than the predators can eat. There is safety in numbers; most individuals in the swarm will survive because a relatively few of their companions have so completely filled the stomachs of the predators that they can eat no more.

The most important predator that feeds on desert locust eggs is a fly of the genus *Stomorhina,* a relative of the well-known bluebottle flies. The female places her eggs in the froth that surrounds a newly laid egg pod. As the maggots make their way down into the pod, they eat some of the eggs and damage others. The nymphs that hatch from the surviving eggs at the bottom of the pod usually die because they cannot reach the soil surface through the mass of rotting eggs above them. Locust eggs are also attacked by other flies and beetle larvae and, on occasion, even by birds. Birds seldom find egg pods, but on at least one occasion in northern Arabia, hoopoes and wagtails ate many pods that had been exposed by wind erosion.

Ashall and Ellis saw no evidence of disease in dead desert locust hoppers—although other investigators have—and reported that they are killed mainly by cannibalistic companions and predators. Helpless nymphs that have just molted their hard skins and are still soft are often eaten by nymphs that molted earlier and have already hardened their new skins and become mobile. Under some circumstances, when food is becoming scarce for example, cannibalism can promote the survival of a swarm by nourishing its surviving members.

Nymphs of the desert locust are also eaten by scorpions, spiders, ants, other arthropods, lizards, snakes, birds of many species, aardvarks, jackals, hyenas, hedgehogs, various rodents, and people. Birds are the most important of the predators of hoppers, and many species of them are involved, but the most important are wagtails, kites, ibises, storks, and the Egyptian vulture. Observations in the field showed that a kite may eat as many as 1,200 hoppers in a day and a stork about 2,000 per day.

Flying swarms of adult locusts may be followed for days by birds, including vultures, storks, and many others that eat thousands or even millions of locusts. More sedentary creatures that live in an area where a swarm alights also take advantage of the opportunity to gorge themselves.

But to my mind, the most intriguing of the creatures that prey on adult desert locusts are large solitary wasps that themselves form groups to follow locust swarms. In a fascinating book, *Insect Migration,* C. B. Williams described his experience with these wonderful creatures. These wasps, *Sphex aegyptius,* have no common name. They belong to a

group of nonsocial wasps that provide food for their larvae by stocking their nest burrows with prey that they paralyze with their venomous sting. Different species specialize on different prey, but quite a few of them, including the *Sphex* that Williams observed, stock their nests with grasshoppers.

In February of 1929, he saw a great swarm of desert locusts settle on the ground. Within 15 minutes he noticed dozens of the large, black *Sphex* running about on the ground. They were very numerous. His two assistants caught 168 of them "within an hour on an area of only a few square yards." As Williams wrote, "immediately on arrival the [wasps] began to burrow, and a little later were dragging paralysed locusts along the ground and into their burrows. This continued all that day till dusk, and started again the following morning shortly after 7 A.M. Between 1 and 2 P.M. on the second day the locusts on the ground began to fly, and a rapid flight to the north or north-east set in: at 2:15 only three live and one dead *Sphex* were seen . . . where two hours previously there had been thousands." Although Williams does not say so, the wasps presumably left to follow the locusts. He goes on to say that the wasps "departed in such a hurry that they left hundreds of open burrows, many half finished; and paralysed locusts were lying about in dozens, some just alongside the burrows in which they should have been interred."

⚘ ⚘ ⚘ A swarm of locusts is awesome and frightening and can be incredibly destructive. Although most insects are harmless, some of the few that attack us or our crops may sometimes, like locusts, be so numerous and destructive that they are a plague upon the land. Let us now turn to some of the ways that people have sought to alleviate plagues of locusts and other insects, some of them eminently practical and effective and others, such as resorting to magic or calling upon a religious organization to condemn the offending insects, are rooted in ignorance and, as far as anyone has been able to determine, are totally ineffective.

People and Insect Plagues

*Immature and adult chinch bugs sucking sap
from a cornstalk*

✸ ✸ ✸　Humanity would probably not survive if all or only certain critically important insects were to disappear from the earth. Insects are necessary and indispensable components of virtually all the ecosystems in which we live and which provide the food we eat. Nevertheless, some few insects are indisputably detrimental to humans. Most destructive insects are more or less stealthy and often not obvious to the eye, but some, such as locusts, occur as huge hordes and are all too apparent plagues on the land. The ancient Israelites recorded the depredations of insects, including great plagues of migrating locusts. In the Bible we can read in Joel (1:4) a litany of the devastation wrought by insects: "That which the palmer-worm hath left hath the locust eaten; and that which the locust hath left hath the canker-worm eaten; and that which the canker-worm hath left hath the caterpillar eaten." (Palmer-worms and canker-worms are actually two different kinds of caterpillars.) Some biblical scholars have suggested that the four insects mentioned by Joel might really have been different kinds of locusts or locusts in different stages of growth. That may be. I can't dispute it. But even so, the message remains the same: insects can occur as plagues and they can cause great devastation.

The Bible tells us that plagues have been visited upon humans as divine punishment. Jehovah forced the Egyptians to release the Israelites from slavery by subjecting them to ten great afflictions—ranging from first turning the water of the Nile to blood and finally to killing the first-born son of every Egyptian family (Exodus 7–10). After each of the afflictions except the last one, the pharaoh at first agreed to free the Jews but then hardened his heart and changed his mind. The second affliction was a plague of frogs that erupted from the Nile and covered the land, even crowding into the bed chambers, kneading troughs, and ovens of the enslavers. The third was a plague of lice. The dust turned to lice when Aaron smote the ground with his staff, and the lice were "upon man and upon beast." The fourth was a plague of flies that swarmed over the land and filled the houses of the Egyptians. The

eighth affliction, the last one involving insects, was a great plague of locusts that Jehovah sent to "cover the face of the earth . . . eat every tree that groweth . . . and eat every herb of the land."

Insect plagues have occurred and continue to occur almost everywhere on earth. Swarms of migrating locusts have not been seen in North America since 1879, probably because they have become extinct. But other kinds of locusts still devastate areas of South America, Africa, Europe, Asia, and Australia. Although chinch bugs have seldom been destructive to field crops in recent years, only a few decades ago great hordes of these destructive suckers of sap could be seen in the midwestern United States as they abandoned fields of dry, ripening wheat and marched overland to invade and devastate fields of still succulent corn. In Eurasia, North America, and Africa—at this very moment in Nigeria—great masses of the caterpillars known as armyworms feed in fields of grass or small grain, and when they have eaten everything, they, like chinch bugs, march over the ground as vast and ravenous armies in search of more green food. Gypsy moth caterpillars can be so numerous in many parts of North America that their fecal pellets falling from the trees sound like steady rain. They often strip all the leaves from trees, leaving many square miles of summer woodland looking like a winter landscape. In western North America, hungry hordes of the same wingless katydids, known as Mormon crickets, that threatened the crops planted by the first settlers in the valley of the Great Salt Lake in Utah still march down from the hills to ravage the plants on the cultivated lands in the valleys below.

Our dependence upon cultivated plants makes us very vulnerable to the destructive potential of insects—far more so than we were thousands of years ago before we became agriculturists, while we were still hunter-gatherers who exploited scattered plant and animal resources. Our crops, usually grown in dense stands as a monoculture, are a great convenience for us and at least as great a convenience for insects. Although some insects, such as grasshoppers and armyworms, are generalists that will feed on many different crops, many others are specialists that will feed on only one crop. Both types are benefited by monocultures of plants, but the specialists more so than the generalists, because they don't have to search for widely scattered wild plants of their preferred species. Monocultures of crop plants that cover large areas, often thousands of acres, are a concentrated and virtually unlimited re-source for insects. Thus the population of a pest insect, not limited by the unavailability of food, can quickly increase to astronomical levels and become a true plague. A field of corn or any other cultivated plant is a veritable banquet that we set before the insects, and can be easily and rapidly devoured by them.

⊛ ⊛ ⊛   Before the advent of modern insecticides in the twentieth century, many insects could not be controlled and others were con-trolled only with great difficulty. For example, late in June of some year in the nineteenth or early twentieth century you might have seen a mid-western farmer devoting many days of concentrated effort to construct-ing and maintaining an intricate barrier to block a massive migration of millions of chinch bugs that were abandoning a field of ripening and drying small grain—very likely wheat—to walk to an adjacent field of still growing and succulent corn. This great horde of sap-sucking bugs would soon have destroyed the corn if they had been allowed to com-plete their migration.

In early spring, adult chinch bugs, which survive the winter hidden in ground litter or tufts of grass, fly to fields of wheat or other small grains to lay their eggs. The parent bugs soon die, but their progeny feed on the grain until the plants begin to ripen in June. Then, when most of them are still wingless nymphs, the bugs leave the drying grain and move overland on foot. Many find their way to adjacent or nearby corn fields, where they do great damage. Chinch bugs feed only on grasses, and corn, like the small grains, is a grass. (This migration prob-ably reflects a similar migration that chinch bugs, native to the prairies of the United States, had been making for millennia before the advent of agriculture, a migration from dying annual prairie grasses to still succu-lent large-stemmed perennial prairie grasses.) The chinch bugs produce a second generation on corn, and it is these bugs, the grandchildren of those that survived the previous winter, that will survive the coming winter—at least some of them—to continue the yearly cycle.

Today chinch bugs, although they still sometimes injure lawns, are not often destructive to field crops in the midwest or anywhere else; their numbers have greatly dwindled because soybeans have largely re-

placed small grains in this area, thus disrupting the two-plant feeding economy of these insects.

But earlier in this century, when small grains were still extensively grown in the corn belt, farmers were frequently plagued by this highly destructive pest. Before the development of the modern insecticides, farmers could contend with chinch bugs only by laboriously constructing barriers between small grain fields and corn fields. The barrier began with a single furrow that the farmer, following behind his horse or mule, plowed between the two fields, throwing up a ridge of soil at the side of the furrow toward the corn. The valley of the furrow was then smoothed and firmed by dragging a log or a small, heavy keg of water up and down its length. When the weather was dry, farmers sometimes dragged the furrow long enough to produce a layer of fine dust. Next, post-holes were dug in the furrow at intervals of about 16 feet. Finally, a line of coal-tar creosote was poured along the ridge just below its brow and on the side facing the approaching chinch bugs.

The migrating bugs were trapped in the furrow. Many lost their footing in the dust, and those that managed to climb the ridge were repelled by the creosote at its crest and fell back into the furrow. The furrow became crowded with milling chinch bugs, many of which fell into the post-holes, where the farmers killed them by pouring a little kerosene over them or crushed them by repeatedly slamming a post down into the hole. The barrier was tended for days as repairs were made and as fresh creosote was applied to the ridge. These barriers saved many a corn crop if they were well made and well maintained. As C. M. Packard and two coauthors wrote in a 1937 *Farmer's Bulletin*, "In one instance 9 bushels of bugs were caught along 1/2 mile of creosote barrier in a week, and approximately the same quantity in the same barrier the next week. It was estimated that at least 60 million bugs were caught along this line in a week."

Preventing chinch bugs from invading corn fields became simpler and much less labor-intensive with the development of the modern synthetic organic insecticides. In the unlikely event that a migrating army of chinch bugs should appear today, it could be stopped virtually in its tracks by spraying a barrier strip of insecticide only about 20 feet wide between adjacent fields of corn and small grain. The bugs would be killed by residual insecticide absorbed through their feet as they walked across the barrier strip. (Most of the modern insecticides are contact poisons: they are absorbed through an insect's "skin" and need not be ingested to do their deadly job.) This barrier strip is an exceptionally—almost uniquely—parsimonious and efficient use of an insecticide. A square 100-acre field of corn can be protected by putting insecticide on slightly less than 1 acre of land.

Insecticides have revolutionized insect control. They simplify many control procedures other than stopping a migration of chinch bugs, and make it possible to alleviate some of the insect problems that simply had to be endured before insecticides were discovered. Mormon crickets and grasshoppers can be killed by poison baits made of bran, ground corncobs, oil, and an arsenical compound. The arsenicals, whose insecticidal properties were discovered late in the nineteenth century, were the first insecticides to be widely used in agriculture. They are stomach poisons, and thus do not kill unless they are eaten. Therefore, if they are sprayed on plants, they kill insects that chew on the plant but do not kill aphids and other insects that pierce the plant to suck sap from below the arsenic-bearing surface. The modern synthetic insecticides—most of which, as you read above, are contact poisons—are more versatile and can be sprayed on any surface that the insect contacts. They have a multitude of uses. If applied early enough, they can nip a developing infestation of gypsy moth caterpillars in the bud. Swarms of migratory locusts can be prevented from forming by spraying their breeding grounds with a contact insecticide, or—although this method is less efficient—many locusts have been destroyed by spraying flying swarms from aircraft. Insecticides are used all over the world to combat hundreds of different kinds of insects that are detrimental to humans in many different ways: mosquitoes that suck our blood and transmit diseases, brown planthoppers that can devastate a rice crop in Asia, caterpillars that threaten to damage a field of lettuce in California, and beetles that infest wheat stored in a midwestern grain elevator.

But insecticides are not a panacea. They are a mixed blessing. Some pests cannot be practically controlled by them; hundreds of insects have become resistant to one or more of them; they can raise an uncommon insect to pest status if they kill its parasites and predators but do not kill it; they may poison important pollinating insects; they have caused massive fish kills; and they have nearly wiped out some species of birds.

Worldwide, insecticides kill hundreds of people every year and sicken thousands more. In *Nature Wars*, Mark Winston relates that orchardists who spray their apple trees with Guthion, an insecticide in the same chemical group as the chemical warfare agents known as nerve poisons, are so sickened despite their protective gear that they must often get off their tractors to vomit.

Nevertheless, although insecticides are a far from perfect solution to insect problems, they have their place and we will continue to need them in the foreseeable future. But there is hope that our reliance on these toxins will decrease. The aphorism "know your enemy" is especially applicable to the field of insect control. Insecticides can be most efficiently used only if we understand the seasonal occurrence, behavior, physiology, and ecology of pest insects. Furthermore, understanding the ecology and behavior of pest insects has made it possible to develop such alternatives to insecticides as crop rotation; pest-resistant varieties of crop plants; and biological controls, the encouragement or introduction of parasites or other enemies of pests. And more alternative controls will become apparent to us as we learn more about our insect enemies.

❀ ❀ ❀   Before insecticides there were few options for a timely response to a present or imminent threat from destructive insects. There were biological controls and cultural controls such as deep plowing or rotating crops. Although these were excellent preventive measures that are still in use, they could not then and cannot now be used to alleviate an immediate problem. The depredations of a few pest insects could be lessened by physical means that were usually labor-intensive. An invasion by chinch bugs, armyworms, and a few other insects that migrate on foot could be stopped by physical barriers. Insects that spend the winter hidden in dry grass or ground litter, a minority of the pest species, could be destroyed by burning their winter refuges, but the fire also destroyed useful insects and was often not widespread enough to eliminate the pests as a problem. Children were set to work picking Colorado potato beetles from the plants by hand. There were also largely ineffective mechanical contrivances, among them vacuum devices to suck up various plant-feeding insects and "hopperdozers" that scooped up grasshoppers into a trough as the dozer was drawn through the field by horses.

But in those early days most pest insects, migratory locusts among them, could not be stopped, because there were no insecticides to kill them or because so little was known about their ecology and habits that it was impossible to develop noninsecticidal strategies for their control. People were more often than not helpless when faced by an insect attack. The arrival of a large swarm of migratory locusts, for example, was and still can be an unmitigated disaster. Crop plants and almost everything else that is green are destroyed. Even the tender bark of trees and the straw in brooms may be eaten. Attempts to kill enough of the invading locusts to save a crop were futile. Farmers burned them with torches and smashed them with brooms and spades, but the locusts kept coming. It was like trying to drain a river with a bucket.

In the 1937 film version of Pearl Buck's *The Good Earth*, a novel set in a community of impoverished farmers in northern China, a great swarm of locusts is about to arrive when the grain is almost ready to harvest. Under the direction of the educated son of the novel's protagonist, the farmers clear a fire-break through the wheat and douse a strip of the dry grain with kerosene. The locusts darken the sky as they arrive. They descend and begin to devour the wheat. The fire barrier is lit and the farmers trample locusts with their feet and crush them with hoes and spades, but the ravenous insects keep coming. It seems that the battle has been lost. But the author saves the day by using a literary device—a *deus ex machina*—a wind that blows the locusts away. Victory! The crop is saved, but it is obvious to the viewer that the efforts of the farmers had had only a trivial effect. It was the wind that saved the day.

✿ ✿ ✿ In the face of their helplessness, people have resorted to magic and pleas for divine intervention to alleviate insect problems. In his *Natural History*, published in 77 c.e., the Roman encyclopedist Pliny the Elder recommended several magical remedies for warding off the attacks of insects. (Pliny's writings were taken seriously well into the Middle Ages of Europe.) For the protection of millet from worms (caterpillars) and sparrows he recommended that "a bramble-frog should be carried at night round the field before the hoeing is done, and then bur-

ied in an earthen vessel in the middle of it . . . The frog, however, must be disinterred before the millet is cut; for if this is neglected, the produce will be bitter." He wrote that caterpillars that infest flower gardens "may be effectually exterminated, if the skull of a beast of burden is set up upon a stake in the garden, care being taken to employ that of a female only. There is a story related, too, that a river crab, hung up in the middle of the garden, is preservative against the attacks of caterpillars."

Pliny's information and recommendations are not original with him. They came from many sources, most of which he did not acknowledge. He was not, to say the least, a critical author, and he passed on some wild and fanciful tales. In all seriousness, he related the fantastic tale that in India there are ants that mine gold and are as big as wolves, and locusts that are 3 feet long. The latter story is echoed by postcards from Texas that show two hunters carrying on their shoulders a pole from which is slung a "Texas grasshopper," a huge creature that appears to be at least 3 feet long.

❈ ❈ ❈   Humans have probably asked deities for help with their problems since earliest times. Twenty thousand or more years ago a Cro-Magnon shaman may have prayed for success in a hunt for mammoths or for relief from a scourge of mosquitoes or biting flies. The ancient peoples of the Middle East, as Isaac Harpaz reminds us, regarded pestiferous insects "as a kind of divine punishment meted out on the sinful. Hence there is nothing to be done about it except meekly submitting to it in penitance, making prayers, offerings, or other rituals as prescribed by the respective religion." The ancient Greeks believed that different gods had domain over different kinds of vermin, and that it was necessary to invoke the help of the appropriate god to prevent or alleviate a pest problem. Zeus, nicknamed the "fly-catcher," held sway over flies; Hercules controlled locusts and caterpillars; and Apollo controlled mice and mildew.

According to the Bible, almost 3,400 years ago Moses pleaded with Jehovah to relieve the Egyptians of the eighth plague, the plague of locusts, because Pharaoh had promised—but once again falsely, as it turned out—to let the Israelite slaves go. "And Jehovah turned an exceeding strong west wind, which took up the locusts, and drove them into the Red Sea; there remained not one locust in all the border of Egypt" (Exodus 10:19). About 400 years later, Solomon, king of ancient Israel, offered, at the dedication of the first temple in Jerusalem, a prayer (I Kings 8:23–53) that included a plea for protection from locusts. Somewhat less than 2,000 years ago Pliny told the story of how the god Jupiter answered the prayers of a community of farmers plagued by locusts by sending a flock of rose-colored starlings (*Sturnus roseus*) that devoured the locusts.

A little over 150 years ago, the first Mormon settlers in the valley of the Great Salt Lake in what is now Utah planted wheat. But their crop was threatened with destruction by a huge horde of wingless katydids, now inappropriately known as Mormon crickets, that crawled down from the surrounding hills. A flock of California gulls, which nest in nearby marshes, came to the rescue and ate the insects. In 1913 the Mormons commemorated what they thought to be their miraculous rescue from starvation by placing a golden statue of a California gull in Temple Square in Salt Lake City. To this day Mormon crickets sometimes come down into the valley and sometimes California gulls come to eat them.

Twenty-seven years after the gulls saved the Mormons' wheat, on May 17 in 1875, the governor of Missouri, C. H. Hardin, issued an official proclamation to promote divine intervention to rid his state of a developing infestation of Rocky Mountain locusts:

Whereas, owing to the failures and losses of crops, much suffering has been endured by many of our people during the past few months, and similar calamities are impending upon large communities, and may possibly extend to the whole state, and if not abated will eventuate in sore distress and famine;

Wherefore, be it known that the 3rd day of June proximo is hereby appointed and set apart as a day of fasting and prayer, that the Almighty God may be invoked to remove from our midst those impending calamities, and to grant instead the blessings of abundance and plenty; and the people and all the officers of the State are hereby requested to desist, during that day, from their usual employments, and to assemble at their places of worship for humble and devout prayer, and to otherwise observe the day as one of fasting and prayer.

As Howard E. Evans wrote in *Life on a Little-Known Planet,* a wonderfully interesting book on natural history, "the people fasted and prayed as proclaimed by the governor, and lo, within a few days the locusts began to leave and die." By the fourth of July the locusts were gone and the country was green and prosperous. But well before he issued his proclamation, the governor had received from the brilliant Charles Valentine Riley, then the Missouri state entomologist, a report, which may have gone unread, predicting that the locusts, which were known not to remain for long in the Mississippi drainage area, would begin to disappear from Missouri in early June.

Riley's remarks on the governor's proclamation, published in the *St. Louis Globe* on May 19, 1875, are exactly to the point:

I deeply and sincerely appreciate the sympathy which our worthy Governor manifests for the suffering people of our western counties, through the proclamation which sets apart the 3d of June as a day of fasting and prayer that the great Author of our being may be invoked to remove impending calamities. Yet, without discussing the question as to the efficacy of prayer in affecting the physical world, no one will for a moment doubt that the supplications of the people will more surely be granted if accompanied by well-directed, energetic work. When, in 1853, Lord Palmerston was besought by the Scotch Presbyterians to appoint a day for national fasting, humiliation and prayer, that the cholera might be averted, he suggested that it would be more beneficial to feed the poor, cleanse the cesspools, ventilate the houses and remove the causes and sources of contagion, which, if allowed to remain, will infallibly breed pestilence in spite of all the prayers and fastings of a united but inactive nation. We are commanded by the best authority to prove our faith by our work. For my part, I would like to see the prayers of the people take on the substantial form of collections, made in the churches throughout the State, for the benefit of the sufferers, and distributed by organized authority; or, what would be still better, the State authorities, if it is in their power, should offer a premium for every bushel of young locusts destroyed. In this way the more destitute of the people in the infested districts would have a strong incentive to destroy the young locusts, and thus avert future injury, and at the same time furnish the means of earning a living until the danger is past. The locusts thus collected and destroyed could be fed to poultry and hogs, buried as manure, or dried, pulverized and sold for the same purpose.

❁ ❁ ❁   During the Dark Ages of Europe religion was very much involved in attempts to eliminate insect plagues, often in ways that seem bizarre today. There were, of course, the usual prayers for divine intervention, a practice that continues to the present day. But sometimes more explicit and direct religious measures were taken. Late in the ninth century c.e., the area around Rome was plagued by grasshoppers. The destruction of millions of them by the peasants did not alleviate the problem. As the story goes, Pope Steven VI prepared huge quantities of holy water and had the infested area sprinkled with it. The grasshoppers immediately disappeared. Sometimes a saintly person drove off pestiferous insects, as did St. Bernard of Clairvaux. As David Bell told the tale in his book on beastly tales, Bernard went to Foigny for the dedication of a new church, but found it "infested by an incredible swarm of flies; and their buzzing and ceaseless flying about was a very great nuisance to those coming in." Since the flies were intolerable and there was nothing else that could be done, Bernard said, "I excommunicate them!" In *The Criminal Prosecution and Capital Punishment of Animals*, E. P. Evans wrote of this event:

> William, Abbot of St. Theodore in Rheims, who records this miraculous event, states that as soon as the execration was uttered, the flies fell to the floor in such quantities that they had to be thrown out with shovels . . . This incident, he adds, was so well known that the cursing of the flies of Foigny became proverbial and formed the subject of a parable.

Evans goes on to say:

> According to the usual account, the malediction was not so drastic in its operation and did not cause the flies to disappear until the next day. The rationalist, whose chill and blighting breath is ever nipping the tender buds of faith, would doubtless suggest that a

sharp and sudden frost may have added to the force and efficacy of the excommunication.

But the church went beyond prayers, exhortations, and the sprinkling of holy water—going so far as to establish formal ecclesiastical courts that put pest insects on trial and condemned them if they were found guilty. The ecclesiastics, however, were on the horns of a dilemma. On the one hand, it was usually assumed that destructive insects or other vermin were sent at the instigation of Satan to pester people. On the other hand, as pointed out by E. P. Evans, they might be "creatures of God and agents of the Almighty for the punishment of sinful man." In the latter case an "effort to exterminate them by natural means would be regarded as a sort of sacrilege, an impious attempt to war upon the Supreme Being and to withstand his designs." In either case, whether the insects were agents of the devil or emissaries from a wrathful God, the opinion of the day was that the only proper and permissible way to find relief from insect plagues was through the offices of the church. As Evans wrote:

> If the insects were instruments of the devil, they might be driven into the sea or banished to some arid region, where they would all miserably perish; if, on the other hand, they were . . . ministers of God, divinely delegated to scourge mankind for the promotion of piety, it would be suitable, after they had fulfilled their mission, to cause them to withdraw from the cultivated fields and to assign them a spot, where they might live in comfort without injury to the inhabitants. The records contain instances of both kinds of treatment.

Despite this ambivalent attitude, there were many trials of locusts, caterpillars, weevils, and other troublesome insects. Roger Swain wrote that by the sixteenth century the method of bringing insects to trial had become formalized and exact, and that a Burgundian lawyer, Bartholomaeus Chasseneus, had written a treatise on the rules for bringing suit against grasshoppers.

Opinions were divided on the validity of excommunicating insects and other nonhuman creatures. Many theologians denounced the idea on grounds that nonhumans could not be communicants of the church and, therefore, could not be excommunicated. For example, in the thirteenth century, Thomas Aquinas argued that no "animal devoid of understanding can commit a fault," and therefore no animal can be reasonably punished by the church. In a similar vein, the attorney for the defense in a trial of weevils that were destroying a vineyard argued that the insects could neither be tried nor excommunicated because they were brute beasts subject only to natural law, not to human or canon law. But the prosecutor took the opposite view, as did some clerics of the day. He argued that God had made insects and other animals to be subordinate and subservient to humans, to be the vassals of humans. Then he concluded that insects, as the subordinates of humans, are subject to excommunication. Others contended that since the lower animals are "satellites of Satan," it is proper to put on them the worst possible curse, which is excommunication. As Evans wrote, it was in the interest of ecclesiastics to go along with insect trials because "it strengthened their influence and extended their authority by subjecting even the caterpillar and the canker-worm to their dominion and control."

In the Tyrol in 1338, an ecclesiastical court tried a devastating population of locusts, found them guilty, and, in Evans' words, instructed the local parish priest to "proceed against them with the sentence of excommunication in accordance with the verdict of the tribunal." Evans goes on to say:

> This he did by the solemn ceremony of "inch of candle," and anathematized them "in the name of the Blessed Trinity, Father, Son, and Holy Ghost." Owing to the sins of the people and their remissness in the matter of tithes the devouring insects resisted for a time the power of the Church, but finally disappeared.

There is no doubt that the locusts died a natural death or moved on of their own volition.

Insects to be prosecuted were often represented by defense counsels. The interplay between the prosecutor, who sought to condemn the insects, and the defense counsel, who fought hard to defend them, reflects the ambivalent medieval attitude toward destructive insects: Were they tormenting agents of the devil or were they punishing emissaries from God? The insects sometimes won the day and were not condemned. They were entreated to leave by prayers; the people were exhorted to

pay their tithes to the church and to mend their sinful ways to relieve a punishing plague; and sometimes the insects were asked to move from the fields they were injuring to other places that were set aside for them. This ambivalence is illustrated by the trial of the weevils that infested the vineyards of St. Julien in France.

The archives of the city of St. Jean-de-Maurienne, near St. Julien, include the original records of legal proceedings against the weevils. In 1545, the wine growers lodged an official complaint against the weevils. In the hearing that followed, the weevils were represented by the procurator Pierre Falcon and the advocate Claude Morel. The wine growers, the plaintiffs, were represented by Pierre Ducol. François Bonnivard, the official who heard the case, did not pass sentence, but instead issued on May 8, 1545, a proclamation:

> In as much as God, the supreme author of all that exists, hath ordained that the earth should bring forth fruits and herbs . . . not solely for the sustenance of rational human beings, but likewise for the preservation and support of insects . . . it would be unbecoming to proceed with rashness and precipitance against the animals now accused and indicted; on the contrary, it would be more fitting for us to have recourse to the mercy of heaven and to implore pardon for our sins.

The weevils eventually disappeared, but in 1587 they reappeared in numbers and were again destructive. On April 13 of that year, the weevils were brought to trial before his most reverent lordship, the prince-bishop of Maurienne. Since the trial was fraught by incessant delays, the inhabitants of St. Julien were called together in a public meeting on June 29, 1587, to consider the propriety of setting aside for the weevils "a place outside of the vineyards of St. Julien where they might obtain sufficient sustenance without devouring and devastating the vines of the said commune." A site was selected and dedicated to the use of the weevils.

But the insects were deaf to the exhortations of the people and did not move to the reservation that had been set aside for them. On July 24, a record of the public meeting was submitted to the court and the court was asked to order the insects, on pain of excommunication, to accept the generous offer that had been made to them and to leave the vir⌐

yards and move to their reservation. The trial was again delayed, and it was not until September 3 that the representative of the weevils responded.

He said that his clients could not accept the offer because the reservation was "sterile and neither sufficiently nor suitably supplied with food" for the weevils. The representative of the wine growers countered that the place set aside for the insects was admirably suited to them, "being full of trees and shrubs of diverse kinds." The court appointed experts to examine the weevil reservation and submit a written report on its adequacy. We will never know if the weevils were ultimately excommunicated, because, as Evans found, the last page of the court records was destroyed by "rats or bugs of some sort."

In the middle of the seventeenth century, the area around Segovia in Spain was plagued by a destructive population of grasshoppers. As William Christian related the story, numerous remedies were tried during a period of 2 years, but all failed. Among them were the public admonitions to the sinners to reform, the saying of novenas, the sprinkling of the area with special holy water from Navare, and even the performance of exorcisms. In 1650 the grasshoppers were brought to trial in the Hieronymite monastery of Santa Maria de Párraces near Segovia. The prosecutors and witnesses were all saints or souls in purgatory who were represented by and spoke through local villagers. The judge, advised by stand-ins for Saint Francis, Saint Jerome, and Saint Lawrence, was Our Lady Saint Mary, who spoke through her representative, the prior of the monastery. After hearing the evidence, she ruled that the grasshoppers would be automatically excommunicated if they did not leave.

The legal prosecution of insects continued until surprisingly recent times—until long after the onset of the Industrial Revolution and the invention of the railroad and telegraph. The most recent prosecution of insects mentioned by Evans occurred at Pozega in Slavonia, eastern Croatia, in 1866 when the region was plagued with locusts. One of the largest of the locusts was seized and tried, found guilty, and then put to death by being thrown into water with anathemas pronounced on it and the whole species.

It is not easy for us to understand how people can conceive of dispelling insects through religious ceremonies or with magic. But in the Mid-

dle Ages people viewed life and their world very differently than we do today. Religion was at the center of people's lives and influenced virtually their every activity and thought. It gave them the rules of civilization that governed their daily lives. But religion was also much more than that to the people of the time, people who had as yet discovered very little about the workings of the natural world. It offered them answers to or protection from the unknown, from that which was not understood and was therefore frightening: a solar eclipse, the appearance of a comet, a sudden outbreak of disease, or even a plague of insects.

At that time the minds of most people were also greatly influenced by superstition. Evans illustrates this uninformed mind-set in a discussion of "bewitched kine." At one time European peasants often penned their cattle for the night in stalls so small that the poor beasts suffered from a lack of fresh air and oxygen and spent the night stamping their hooves and making agitated movements. The peasants attributed this agitated behavior to demonic possession caused by witchcraft. When a veterinarian tried to explain to them that the cattle were suffocating and that the problem could be solved by keeping the windows open, the peasants did not take his advice. However, the peasants did open the windows when he told them that if they were kept open the witches could enter and leave freely during the night and would not cause demons to enter their cattle.

Finding a Mate

*The glowing snares of thousands of Waitoma fly larvae
suspended in a cave; below is the adult*

⚘ ⚘ ⚘   On a sunny January day, I walked in a woodland along the Sangamon River in Illinois, the same woodland in which I had, in the spring of an earlier year, found the aggregation of ladybird beetles. It was a crisply cold day, and the thin blanket of snow squeaked beneath my boots as I stepped along. A large dead tree stood by the side of the trail, big pieces of loose bark still clinging to its trunk. I pulled off a small slab to see what might be hiding under it. There were a spider, a few small beetles, and a miscellany of other small insects, but the creature that first caught my eye was a large yellow and black wasp clinging to the underside of the bark. It was a queen yellowjacket. Yellowjackets are social and live in large colonies during the warm season, but the colonies cannot survive the winter. Until spring, the species is represented only by new, already fertilized queens that are nestled in nooks and crannies, often beneath a flake of bark or in some other protected place. The queen that I found did not move. She was in a state of diapause that would not end until spring, when she would fly off to found a new colony.

I knew that she was prepared to lay fertilized eggs when she awoke in spring. Her spermatheca, an internal storage organ for sperm, was presumably full of semen from males she had mated with the previous autumn. Almost all insects except mayflies have spermathecas, wonderfully useful organs that make it possible for a female to couple only during a short mating season, in some species with only one male, and yet lay fertilized eggs weeks or even years thereafter. These insects, including yellowjackets, can thus mate at the most propitious time for that activity but delay laying eggs until some other time, perhaps in another season, that is more favorable for the survival of their offspring.

Spring is, of course, the best time for a yellowjacket queen to found a new colony. The insects these wasps feed to their larvae are becoming abundant, and a long warm season stretches ahead during which the colony can increase its population. The lone queen feeds and otherwise

cares for the first small batch of larvae she produces. They will become workers, and when they mature, they will relieve the queen of parental duties, and thereafter she will devote herself only to laying eggs. As the season progresses, the colony grows in size, populated by the queen, usually the one member of the colony capable of reproducing, and as many as several thousand workers—every one of them a daughter of the queen—which never mate but can produce males under certain circumstances.

The best time for yellowjackets to mate is late summer and early autumn. By then the colonies have grown large enough to produce a generation of males—the first and only ones of the year—and new queens that, if they are lucky, will survive the coming winter to found a colony in the spring. All of the colonies in an area produce many new queens and males at about the same time. Soon they are abundant. You can see hundreds of them sipping nectar in the patches of goldenrod that blossom so abundantly in late summer or early autumn. Large numbers of males swarm on hilltops or near other conspicuous landmarks, a behavior I discuss in detail later in this chapter. Virgin queens orient to the same sites and there are inseminated by a male or, more likely, several males. The males and the workers die when it becomes cold.

❁ ❁ ❁   The mating success of many insects—and of many other animals, too—depends upon how closely individuals of both sexes coincide in time as well as in space. Yellowjackets and virtually all other animals have evolved strategies that assure the synchronization of males and females. Generally speaking, the more closely the members of a population are synchronized, the more likely an individual is to find a mate. (As you will read in the next chapter, the mayflies, most of which live for only a day, are an extreme example. Millions of them become sexually mature and emerge from the water to fly and mate on the same day.) Many insects and many other animals are synchronized because all members of a population "march to the beat of the same drummer." They usually grow and mature at the same rate under a given temperature regimen, and thus all become sexually mature and ready to mate at about the same time. The members of some species are even more closely synchronized because all respond to the same specific environmental cue that signals the time for mating, such as a particular day length, the phase of the moon, or the accumulation of a critical number of degree days, such as the "heating degree days" your television weather person keeps track of.

Cecropia moths, one species of the saturniids that are sometimes known as giant American silkworms, are an interesting and more than usually complex example. Cecropias have become a part of the suburban ecosystem in the eastern half of the United States. The big green caterpillars with orange and yellow tubercles on their backs feed on the leaves of some shade trees and ornamental shrubs, but they are never numerous enough to do significant damage, and despite their fearsome appearance, they are harmless. The colorful moths, with their 5-inch wingspans, are as attractive as any butterfly but are not often seen, because they are strictly nocturnal and live for only a few days. (They cannot feed because they have only vestigial mouthparts, but their bodies contain enough fat to provide plenty of energy for mating and the dispersal of eggs.) More commonly seen are the large silken cocoons that house the pupae in winter and that the caterpillars spin in late summer on a twig of a tree or, more often, among the stems of a shrub. Some cocoons, especially those in trees, are easy to see in winter after the leaves have fallen.

The caterpillar will molt to the pupal stage shortly after completing the cocoon, and after spending the winter in diapause, will metamorphose to the adult stage and wriggle out of the cocoon to mate and, if it is a female, to lay eggs. If you collect a few of these cocoons in autumn or winter and keep them at ambient outdoor temperatures, as on a screened porch, the moths will emerge from their cocoons within a few days of each other in spring and at the same time as the wild moths. A virgin female will, shortly before dawn, release a sex-attractant pheromone that will attract wild males—often dozens of them—to your porch screens.

The timing of the moth's emergence is critical. The eggs must be laid late enough in spring so that the leaves of deciduous trees and shrubs have budded out and are available as food for the newly hatched caterpillars, but early enough so that the slow-growing caterpillars have time

enough to finish growing, spin cocoons, and molt to the pupal stage before the onset of cold weather in the autumn.

Because the lives of cecropia adults are short, it would seem to be obviously advantageous for them to emerge from their cocoons as a synchronous group and thereby improve the odds that they will coincide with members of the opposite sex, enabling them to find mates and become parents. But surprisingly, as James Sternburg and I found, a population of adult cecropias, unlike populations of most adult insects, emerges in two shifts separated by almost a month. At the latitude of Urbana, Illinois, a few adults, usually no more than 10 or 15 percent of the overwintering population, emerge as a synchronized group during the last two weeks of May. Many more, the rest of the overwintering population, emerge as a synchronized group about a month later, during the last two weeks of June. Although the emergence dates differ with the latitude, cecropia moths are present as two distinct groups, probably throughout all or most of their range.

Although this division of the population seems to go against the evolutionary grain, some insects other than cecropias have similarly divided emergences. Therefore, unless natural selection has run amok, it would seem that a division of forces can somehow enhance fitness, even though it decreases the size of the pool of mates available to an individual. The advantage of emerging as two separate groups is succinctly expressed by the aphorism that warns against putting all the eggs in one basket. This makes sense only if all cecropia females produce both early- and late-emerging progeny, and as James Sternburg and I found in further experiments, they do. Thus if the individuals of one of the two groups produced by a mother do not survive, their siblings in the other group may survive to pass their parents' genes on to their offspring, the mother's grandchildren. If, for example, the offspring of the late-emerging group are killed by a summer drought, the offspring of the early group may survive. If, on the other hand, the offspring of the early group are killed by a late frost, those of the late group may survive.

Some cecropia moths emerge early and others emerge later because of a genetically fixed difference in the environmental factors required to terminate their pupal diapause and thus initiate the course of development that will transform them into adults. As you read earlier, the pupae of neither group will begin to develop until they have experienced a sufficient amount of winter chilling. The amount of chilling required is the same for both groups, but from there on the early and late groups differ. Pupae destined to become early-emerging moths are opportunists that begin to develop in response to the least bit of warmth immediately after they have experienced the requisite amount of chilling. They begin to develop at the end of March, but because of the cool weather of early spring, they develop slowly and do not become adults until late May, more than 7 weeks later. By contrast, the pupae destined to become late-emerging moths cannot be triggered to begin developing in response to warm temperatures until after they have experienced a rather long preparatory period of warmth after the period of chilling. They do not develop during this preparatory period. It simply primes them to begin developing after it has had its effect. Not until the first days of June, after most of the early moths have emerged, do these late pupae begin to develop. Because of the warm temperatures of late spring, they develop rapidly and become moths in late June, after only about 3 weeks have passed.

⚙ ⚙ ⚙ Among the insects that rely on specific environmental cues to synchronize their mating activities are certain ants. Bert Hölldobler found that the mating flights of four different species of harvester ants in the arid area of southeastern Arizona were triggered by the onset of the rainy season. All of the colonies in an area released large swarms of males and reproductive females within a few days after the rains began. One of the factors that prevents hybrid matings between the four species is that each species swarms at a different time of day, one between 10:00 A.M. and 11:00 A.M., one between 11:00 A.M. and 1:00 P.M., another between 3:30 P.M. and 5:00 P.M., and the last between 4:30 P.M. and 6:00 P.M. Another ant, *Lasius neoniger*, not a harvester, discussed by Hölldobler and Edward O. Wilson, occurs in the eastern United States and responds even more precisely to environmental cues. All of the colonies of *Lasius* in an area "explosively" release nuptial swarms on a warm and humid day in late August or early September within 24 hours after a moderate or heavy rainfall. In *Journey to the Ants*, Höll-

dobler and Wilson refer to this species as the "Labor Day ant," because of the time of year at which it swarms.

☼ ☼ ☼   But the most precise responses to environmental cues are seen among creatures that dwell in the seas, many of whose mating times are controlled by the tides, as are those of the horseshoe crabs that you met earlier. The timing of the daily tides is determined by the position of the moon with respect to the earth. The amplitude of the tides is controlled by the phases of the moon. The spring tides, the two highest of the lunar month and also the lowest, occur at the time of the new or the full moon, when sun and moon are aligned and combine their gravitational pull on the waters of the seas.

Very few insects occur in the seas, but among them is an amazing little fly known as the one-hour midge, because the adult life span is only from a half to 2 or 3 hours. As both May Berenbaum and Dietrich Neumann have reported, mating and egg laying by these midges are timed by the phases of the moon as they control the tides. The larvae of the midges live and feed on the bottom in the deeper waters of the intertidal zone of rocky shores along the Atlantic and North Sea shores of Europe. The adults lay their eggs in the larval habitat and can do so only when the tide recedes enough to uncover the bottom where the larvae will live, which happens only during the lowest ebb of a spring tide. It is at this time that the short-lived adults emerge as a closely synchronized group to mate and lay their eggs.

The date and hour on which the silvery little fish known as grunion come together in reproductive schools is also determined by the phases of the moon, at least inasmuch as they control the tide. As Boyd Walker described in an article with beautiful photographs by Joseph Brauner, grunion—which occur only along the Pacific coasts of southern California and northern Baja California—mass together to spawn on sandy beaches at night, but only on the first three or four nights of a spring tide, and then only during the first 3 hours following maximum high water. They are among the very few fish that leave the water to spawn. Intent only on reproduction, they come ashore with a wave and strand themselves on the wet sand that is exposed as the wave recedes. Each female rapidly wriggles her body tail-first down into the watery sand

and lays her eggs as one or more eager males spill their semen on the sand beside her. The next wave washes the grunion back into the sea. The temporarily stranded grunion are safe from fish-eating gulls, which are active only during the day, but are prey to crowds of people with flashlights who gather these delicious fish from the sand. The eggs, safe from aquatic predators, develop in their moist nests above the waterline and hatch about 10 days later, when the next spring tide washes them out of the sand.

The fantastic reproductive behavior of the marine palolo worm is even more closely timed and synchronized by the phase of the moon— probably about as closely as possible. These segmented marine worms—relatives of the earthworms and thereby distantly related to insects and other arthropods—practice external fertilization, as do many other animals, among them oysters, horseshoe crabs, and grunion. Not only do palolos not copulate, but males and females don't even cuddle close, as do horseshoe crabs and grunion, when they release their sex cells in unison. Like oysters, the palolos simply release sperm and eggs into the water and leave it up to the free-swimming sperm to find and fertilize eggs. The odds are obviously vastly against any one sex cell but are improved because the palolos in one area release their sex cells at the same time. Off the coast of Bermuda, for example, the palolos swarm exactly three days after the full moon and about an hour after sunset. At that time, the rear halves of the bodies of both sexes break away from the front halves and squirm to the surface of the water, where they release their eggs or sperm at virtually the same moment.

⚙ ⚙ ⚙   Insects that attain adulthood in the same place and at the same time can easily find mates. But, broadly speaking, there are only two ways in which other male and female insects or other animals can get together. They can go to a place where members of the opposite sex can be found, like people visiting a singles bar or teenage boys in the 1950s standing on a corner waiting for girls to come along. The other alternative is to send signals that will attract mates from a distance. The pheromones and other sexual signals that people emit work only from close up, but if we stretch our minds a bit, we can compare the long-distance signal of an insect to an ad placed in the personals section of a

newspaper. I'll come back to the "singles bar" concept later, but first let us consider the long-range signals that insects use.

The only senses that can perceive signals from a distance are vision, hearing, and smell. All are used by insects to attract mates from a distance. At night fireflies flash bursts of light that can be seen by other fireflies from a long way off; male crickets, grasshoppers, katydids, and cicadas are well known for the songs, beautifully described by the late Vincent Dethier, that they sing to attract females. Many thousands of different kinds of insects, usually females but sometimes males, use odors to attract mates from a distance. These sex-attractant pheromones are emitted by various kinds of cockroaches, termites, scale insects, beetles, flies, bees, wasps, and moths.

The great majority of insects that use long-distance signals are solitary suitors. A single male cricket sings from the mouth of his burrow; a female firefly sits alone on a sprig of grass as she flashes in a code that will attract males of her species; just before dawn, a female cecropia moth releases a sex-attractant pheromone that will drift down wind and may attract a male from a mile or more away; a lone male boll weevil sits on a cotton boll as he releases a pheromone to attract a female. But some insects form signal-sending groups and thereby multiply the strength of their signals. Among them are bark beetles that emit sex-attractant pheromones, fireflies that gather by the hundreds or thousands to flash in unison from the same tree, and male periodical cicadas that, as you will read later, join together in a tree by the hundreds or thousands to form loud, droning choruses.

❀ ❀ ❀ In 1937, John Buck reported observations that he and other members of the Seventh Botanical Expedition of the Johns Hopkins University had made of a species of firefly, really a kind of beetle, on the island of Jamaica in what was then the British West Indies.

In front of the expedition's laboratory at Chestervale, in the Blue Mountains of Jamaica, there was a thatch palm which bore below its whorl of leaves an inflorescence a meter in diameter. For about a week in June and again a month later, this inflorescence was transformed nightly into a sphere of seething flame by the flashes of thousands of fireflies which gathered there. Later, other displays were discovered, particularly on two large acacia trees overhanging the Clyde Valley which harboured such prodigious swarms of fireflies that the nebulous glow was visible half a mile away.

All the fireflies on these trees were of one species, *Photinus pallens*, and the females outnumbered the males in the ratio of 4:3. Each firefly flashed regularly about twice a second while walking along the twigs, and entirely independently of any other individual. There was no sign of synchronism or of response between any individuals or between different trees. The flashing was not inhibited by heavy rain, by lightning, or by the beam of a powerful flashlight, but did not occur on moonlight nights. It continued from about 8 P.M. until 3 A.M., and during dull days many of the fireflies remained in the trees all day.

These brightly glowing aggregations seem to function as beacons that attract both males and females from a distance. Buck found that he could artificially induce the formation of a glowing aggregation with a flashlight. When the beam shone onto a bush or the grass, individuals of both sexes soon flew in from all directions, alighted on the illuminated spot, and flashed continually, with only brief interruptions between flashes, but did not flash synchronously, and thus they produced a continuous glow. He found many mating couples in natural aggregations, and concluded, quite correctly, that the function of this cooperative light display is to bring together males and females for the purpose of mating. He discerned no signaling between the sexes in an aggregation, and concluded that they find each other by accidental contact as they wander about among their companions. This behavior is very different from that of solitary fireflies, such as those that we commonly see in North America. The latter find each other through coded signals that males and females flash alternately. A lone female perched on a blade of grass flashes her signal and is answered by flashes from a flying male who knows the species code. She replies and he comes closer and flashes again. Finally, this "conversation" conducted by alternately flashing lights brings the two together.

Gerrit Miller reported that another Jamaican firefly, aptly named *Photinus synchronans*, sometimes forms small flying groups of 20 to 40

individuals that may flash in perfect synchrony for 2 or 3 minutes. But by far the most spectacular and the most famous of the synchronously flashing fireflies are several species that occur from India to southeast Asia and in the East Indies from the Philippines and Borneo east to New Guinea. The best known of these species is *Pteroptyx malaccae,* members of which are the performers in the celebrated "firefly trees" of Thailand.

In a 1938 article in the *Quarterly Review of Biology,* John Buck quoted a translation of Engelbert Kaempfer's remarks on the firefly trees that he saw as he traveled down the Meinam River from Bangkok in the early eighteenth century. Kaempfer's account is probably the first published report of these insects.

> The Glowworms (*Cicindelae*) represent another shew, which settle on some Trees, like a fiery cloud, with this surprising circumstance, that a whole swarm of these Insects, having taken possession of one Tree, and spread themselves over its branches, sometimes hide their Light all at once, and a moment after make it appear again with the utmost regularity and exactness, as if they were in perpetual Systole and Diastole [the alternating contractions and relaxations of the heart].

The light from these trees, produced by tens of thousands of individuals, is visible from far away. Firefly trees are among the most noticeable and amazing of insect phenomena.

The fireflies have their favorite trees, and may occupy the same tree for months or even years. They remain in the tree during the day, when they are not signaling, and thus, as John and Elizabeth Buck wrote in 1976, "each evening's display is begun by these residents rather than by fireflies gathering afresh." The relative permanence of these aggregations is suggested by an old report that Malay rivermen use firefly trees as navigation markers when they travel a river at night, a report corroborated, Buck writes, by an observer who studied the same firefly tree near Singapore over a period of 5 years.

Although both sexes have light-producing organs and both are present in the trees, it is only the males that flash so brightly and in virtually perfect synchrony. The males' display attracts both sexes to the tree. The flashing males maintain a remarkably steady rhythm. The interval between flashes varies with the species, ranging from about a half second to as much as 3 seconds. The common species of Thailand flashes about once every 500 milliseconds (about once every half second), and no member of the group is ever out of phase with the others by more than 20 milliseconds. At 88°F, the exact interval between flashes is 560 milliseconds. Flashing in synchrony results in an intermittent but brighter light than the continuous but dimmer glow that would result if the insects did not flash in unison.

The existence of firefly trees brings to mind two interesting questions: How could this cooperative and seemingly altruistic behavior have evolved? How do fireflies benefit from pooling their signals?

Let us consider the second question first. Many fireflies that live in relatively open areas, such as those that we see in North America, are loners that, as you already know, find mates by means of a "conversation" of flashes between the sexes. But such one-on-one contact will rarely work in the densely forested, jungle-like areas of Asia in which firefly trees occur. In such dense cover the flash of a lone individual cannot be seen from a distance; it is bound to be blocked by a nearby branch or leaf. But even in a dense forest, the glow of a large and brightly shining firefly tree flashing on and off can be seen from far away. (The blinking of a firefly tree that happens to be on the bank of a river can be seen from a boat that is miles away.) This remarkable display attracts thousands of fireflies from a large area surrounding the tree.

Mating couples are commonly seen in firefly trees. If the tree is shaken, fireflies will rain down and many copulating pairs will be among them. It looks as if the main or perhaps the only function of firefly trees is to bring the sexes together. Consider that the life of any animal is governed by three basic imperatives. It must eat and grow. It must avoid becoming a meal for another animal. And it must culminate its life by reproducing. The tree offers little or no food to the predaceous fireflies. Their glow might have the secondary effect of warning predators that fireflies are distasteful and not edible, but the flash of a lone firefly accomplishes the same end. Thus it seems an inescapable conclusion that the function of these brightly glowing aggregations is to promote sexual encounters, the first step of reproduction.

Although females do not join in the synchronous signaling, during the dark intervals between the males' flashes their dim flashes and glows can be seen. They are probably signaling to males to inform them

that a willing sexual partner is nearby. After the females have been inseminated—probably by several males—they leave the tree to get on with the all-important business of distributing their eggs in micro-habitats where their predaceous larvae can find prey and survive. It appears that departing females are continually replaced by newly arriving females.

✦ ✦ ✦   To return to the first question, how such close cooperation could have evolved has been debated. Some have said that male fireflies gather together and flash synchronously for the "common good," an anthropomorphic but reasonable shorthand expression of what is going on. But some biologists have argued that the males could not possibly flash for the "common good," because the individual, the object of natural selection, could not benefit. They claim that "cheating" males that do not flash would benefit more than males that do flash, because they would conserve the energy the other males devoted to flashing. Therefore, it is argued, they would be more likely than "honest" males to survive and reproduce, and males with the genetic tendency to cheat would eventually replace honest males that flash. Thus, these biologists conclude, this cooperative behavior would eventually be eliminated. But group flashing has been going on for ages and shows no sign of diminishing. The critics try to explain this by postulating that something other than simply joining together for the common good must be going on, something that benefits the individual male but not other males.

I believe that they are wrong, because they have ignored the obvious. This cooperative system works because each male that belongs to the common group has as good a chance of inseminating a female as does any other male in the group. The "common good" is the individual's good! The more males that flash and the brighter the signal, the more females that will be attracted. After all, a male that tried to go it alone would not be likely to find a female in the dense forest. I know of no data for fireflies, but as you will read later, there is good reason to believe that a male periodical cicada that joins a large singing chorus is more likely to inseminate a female than is a male that sings alone. Pooling their signals for the "common good" is as likely to work for fireflies as for periodical cicadas.

An experiment that has not yet been done, simple in concept but probably difficult to do, might shed some light on this debate by providing a measure of the relative attractiveness to females of large and small groups of flashing males. If a large group with its brighter flash signal attracts more females *per male* in the group, we can conclude that cooperation between males is beneficial to the individual male. The experiment would consist of placing large and small groups of caged males in a natural habitat and counting how many females come to each cage over a period of time. Males that are attracted to the cages would have to be removed so that the size of the group does not increase.

⚙ ⚙ ⚙ Unlike the fireflies, which are actually beetles, Waitomo flies, which also produce biological light, really are flies and have only one pair of wings, as do all of the true flies. They are named for the caves in New Zealand that became a tourist attraction because of the beauty of the light display produced by huge aggregations of these little flies. After more than six decades of experience with them, the New Zealand entomologist G. V. Hudson wrote of this insect:

> Its extraordinary abundance and spectacular appearance, in certain parts of the Waitomo Caves, have much impressed many visitors to that famous locality, and created an impression in the popular mind unequalled by any other member of our insect fauna. The weird and uncanny places that this larva inhabits:—caves, abandoned mining tunnels; deep, dark, dripping, forest-clad ravines and the like, where it spins its sticky webs and hangs therein, using its brilliant light to attract flies and other small organisms on which it feeds, is probably unparalleled by any other member of the insect world.

The larvae, pupae, and adults of Waitomo flies, sometimes called New Zealand glowworms, are all luminescent, as are the drops of thick sticky goo that the larvae produce to attract and entangle the tiny midges and other insects they eat. Each larva constructs its own snare, a complex affair consisting of a strong, horizontal tube of silk and mucus from which, as Hudson wrote, may hang as many as 50 threads, each with glowing blobs of the sticky goo all along its length at intervals of about a

tenth of an inch, giving it the appearance of a string of shiny beads. The larva forms these threads and lets them drop as it sits on its horizontal support. In the still air of a cave these vertical threads are often as much as a foot long, and occasionally even longer. In places where they are exposed to wind, they are usually no more than an inch or two long. If something is caught on one of its "fishing lines," the hungry larva may retrieve its catch by pulling up the line, a maneuver described by Hudson, or it may climb down the line to suck the juices from its catch, as reported by J. B. Gatenby.

When a larva is ready to pupate, it clears away its hanging lines and suspends itself from its horizontal support by a few threads of silk. Since the adults do not feed, they live for only a brief period, as V. Benno Meyer-Rochow and Eisuke Eguchi reported, and their only functions are to mate and lay eggs.

Females usually mate immediately after they emerge from their pupal skins. The males, which probably keep a watchful eye on the many pupae in their aggregation, anticipate the imminent emergence of a female by landing on her pupa, attracted by its glow, which is brighter than that of a male pupa and is at its brightest just before she emerges. As she squirms out of her pupal skin, several males are likely to be clinging to it. They jostle for position, and one of them—probably the most agile jostler—will inseminate her. But mating sometimes occurs after females have flown away from their pupal skins.

✿ ✿ ✿   Some stinging or toxic insects that warn away predators by advertising their noxiousness with bright colors make themselves even more apparent to the eye—and sometimes to the "nose"—by forming aggregations. These assemblages, depending upon the species, can include anywhere from just a few to hundreds of individuals that may be so closely packed that they touch "shoulder to shoulder." According to Hugh Cott, immatures—nymphs and larvae—are more likely than adults to form such defensive groups. But defensive aggregations of adults do occur and may double as mating assemblages, as is the case with the lycid beetles observed by Tom Eisner and Fotis Kafatos.

Lycids, also known as net-winged beetles, are toxic, and various lizards, birds, mammals, and even some invertebrate predators such as wind scorpions, praying mantises, and ants refuse to eat them. All lycids are warningly colored, some with yellow or a combination of yellow and black. They are mimicked in appearance by a lot of perfectly edible insects, including many different kinds of moths, whose resemblance to the lycids often deters predators.

In nature, lycids are usually found in groups. At the American Museum of Natural History's Southwestern Research Station near Portal, Arizona, Eisner and Kafatos found that in a small patch of white clover there were over 3,000 lycids in a number of dense clusters distributed throughout the patch. To determine how the lycids form aggregations, they staked out four mesh bags containing 150 lycids each, only males in two of them and only females in the other two. The two bags that contained females attracted a total of only 15 lycids, but the two that contained males attracted 294, both males and females. Since the mesh bags eliminated visual cues and the beetles made no sound, the incoming beetles had obviously responded to an odorous pheromone.

As Eisner and Kafatos wrote, "the lycid attractant, aside from its obvious function in [keeping the warningly colored population] densely congregated, also serves appropriately in bringing together the sexes preparatory to mating." Mixed lots of males and females in observation enclosures formed individual mating pairs or small clusters of pairs. Females in enclosures without males were not attracted to each other and remained isolated.

❁ ❁ ❁   The Douglas fir beetle, a bark beetle of western North America, lives, feeds, and breeds just under the bark of a coniferous tree. Like the southern pine beetle you met earlier, it usually attacks sick or injured trees, but can successfully colonize a healthy tree by means of a mass attack involving thousands of flying beetles, enough of them to stop the resin flow and other biochemical defenses of the tree. The attack is coordinated by the behavior of the successful colonizers. The mass attack functions not only to subdue the defenses of the tree but also to bring the sexes together for mating.

The pioneering onslaught is initiated by females that burrow into the bark and then release a pheromone that attracts both males and females in the ratio of two males to one female. The pheromone is contained in

their frass, a mixture of feces and sawdust. During this first phase of the invasion, which lasts for only a few hours, the sex ratio of the established attackers is about two females to one male. These pioneers attract a massive invasion of bark beetles in the ratio of two males to each female, but, as J. A. Rudinsky wrote, the mass attack soon ends abruptly. No more bark beetles are attracted to the tree.

The very sudden halt to the arrival of new beetles was a puzzle, since a female's frass alone may be attractive to other Douglas fir beetles for several days. How, then, Rudinsky wondered, was such an abrupt halt possible? Apparently it was due to some kind of interaction between the males and the females, since, Rudinsky knew, logs experimentally manipulated to contain only pairs of males and females, but no unmated females, did not attract new colonizers.

Like many other bark beetles, the Douglas fir beetle can make squeaky sounds by stridulating, by rubbing together two body parts, in their case one wing cover against the other. Rudinsky wondered what role these sounds, whose behavioral functions were at the time virtually unknown, might play in the mating of these beetles. He placed a log artificially infested with females in an outdoor cage, the entrance to each female's burrow covered with a screen that kept other beetles out but allowed the females to expel their frass. Beetles flew in and landed on the walls of the outer cage in the ratio of two males to each female, the same ratio that is seen when pioneer beetles first attack a tree. Then Rudinsky placed a male beetle on the screened cover of every female's burrow. The males stridulated and no more wild male beetles arrived at the cage. When Rudinsky removed the stridulating males, wild males were again attracted to the walls of the cage. Rudinsky was able to turn the ability of the females to attract males on and off by alternately removing and replacing the stridulating males at the screened entrances to the females' burrows.

When Rudinsky silenced males by removing their wing covers, he was able to prove that it is the sound of the male that stops the females from attracting males. In this experiment, females continued to attract males although silent males were at the entrance to their burrows. But when Rudinsky removed the screens to let the silent males join the females in their burrows, the females no longer attracted new males. This led him to believe that the male Douglas fir beetles might have been producing a close-range pheromone that affected the females. When he removed the females from their burrows, the residual pheromone content of their frass continued to attract new males, despite the presence in the burrows of males that were still able to stridulate and produce their own pheromone. Rudinsky concluded from this that females with males produce, in response to the males' pheromone, a second pheromone of their own that masks the effect of their attractant pheromone.

But why did this complicated behavior evolve? The answer is twofold. It serves the male by preventing his mate from being inseminated by rival males. And it serves both male and female by preventing the tree from becoming so crowded with competing bark beetles that the Douglas fir beetles' reproductive efforts would be endangered.

❃ ❃ ❃   Not all insects attract mates from a distance. The other way of finding a mate is to go to a place—like a singles bar—where members of the opposite sex are likely to be. Some male insects lurk at or near one of the resources required by females. All females need a place to lay their eggs, and most of them need a source of food. Males of the tsetse flies, the insects that transmit sleeping sickness in Africa, wait on or near the large animals of the savannah from which females suck blood. Certain male fruit flies stake out and defend against other males a fruit that egg-laying females are likely to visit. The males of some solitary bees lurk near flowers that females may come to for nectar. Some of the colorful hover flies, also known as flower flies, have it both ways. In the morning, males patrol flowering shrubs and pounce on females sucking nectar from blossoms, and in the afternoon both sexes retreat to a shady woodland, where the males, in the hope of inseminating another female, wait at the rot cavities in trees in which females lay their eggs.

Some male insects take a less direct approach. They wait at a place that females might pass—like those teenage boys that stand on street corners waiting for girls to come by. Some kinds of flies take "waiting stations" on branches, leaves, or the trunks of trees. When an insect of about the right size flies by, the waiting male dashes out to intercept it and briefly grapples with it. If his close-range perception, probably chemical, identifies it as a female of his own species, all is well and the

two go off to mate. Otherwise the male goes back to his station and the startled object of his unwanted attention beats a hasty retreat. If you see a male on his waiting station, you can make him go into action by flipping some small object of about his own size, such as a tiny bit of bark, through his field of view.

On a sunny day in late May 1964, fellow entomologist Bill Downes and I climbed to the top of the highest hill in Sand Ridge State Forest in Illinois, the hill on which the fire tower then stood. Here and there a flower fly hovered just beneath the lowest branches of the small trees that grew at the foot of the tower. They were amazingly adept at maintaining their positions, occasionally drifting a few inches off the mark, but always returning to their original place. I decided, as had J. Anthony Downes long before me, that they must be staying in place by keeping a visual fix on some landmark, perhaps a particular leaf above them or a dapple of sunlight on the ground below them. Occasionally, two flies drifted too close together, and one dashed out to grapple with the other but immediately returned to its original position. When I collected about 20 of these flies, I made two interesting discoveries: they were all males and all belonged to a species (*Eupeodes volucris*) that I had found to be very rare elsewhere.

This led me to two conclusions: First, the males were taking "aerial waiting stations" as they waited for females to appear. Second, since there were many on the hilltop and they were rare elsewhere, they were "hilltopping," as entomologists put it—rendezvousing at an outstanding feature of the topography.

In his report of the first systematic observations of the phenomenon later to be known as hilltopping, John Chapman wrote, "On a warm day in July 1951 [I] climbed Squaw Peak, a rather isolated and rocky summit, 7,996 feet in elevation, some 25 miles northwest of Missoula, Montana. At the summit surprisingly large numbers of insects, particularly various kinds of Diptera [flies] were seen rapidly flying about or resting on the bare rocks." He collected a sample of 351 specimens of these insects. They belonged to eight different families and included 27 sawflies (relatives of the bees and wasps), 25 parasitic wasps, and 299 flies of six different families. The sex ratio of his sample was astonishingly skewed in favor of males. He had collected 322 males and only 19 females, 8 of the latter coupled with males.

This concentration of insects on the very summit of the peak puzzled Chapman for several reasons. The summit was barren and seemingly offered no resources to the insects—no water, no food, and no place to lay eggs. Most of the summit-frequenting insects were at the very top of the rocky summit, within an area that extended no more than 8 feet below its highest point. There were very few of them below that level. And, finally, there was the sex ratio that was so amazingly skewed toward males.

A few other entomologists had observed summit-frequenting insects in various parts of the world, including Harold Dodge and John Seago, whose observations in Georgia were published back to back with Chapman's article in the same journal in 1954. What to make of these observations? What reason could there be for large numbers of insects to gather in often barren places with no discernible resources? The answer arrived at by Chapman and the other observers is, of course, that there *is* a resource, members of the opposite sex. These insects, some of which otherwise occur only as small populations of widely scattered individuals, were gathering at a central meeting place (comparable to a singles bar), guided there by their genetically determined propensity to orient to mountains, high hills, or other outstanding topographic features.

Both sexes come to these places, presumably in equal numbers. But males outnumber females because they tend to stay on the summit, even after a successful mating, in the hope of finding another receptive female. Females are not numerous because they leave shortly after they mate to get on with their highest priority: distributing their eggs in places, often few and far between, that will favor the survival of their offspring. One of the most interesting insects in this regard is the female deer bot fly, which lays eggs on deer, in whose nasal cavities and sinuses her parasitic larvae live and grow. This is one of the flies that Chapman collected on Squaw Peak.

❀ ❀ ❀ In 1927, Charles H. T. Townsend, an entomologist best known for his work on the classification of flies, published an article in which he reported seeing deer bot flies on a 12,000 foot mountain peak in New Mexico. He did not wonder why these flies were on a barren peak where no deer were present. If he had, he might have discovered hilltopping long before Chapman. Instead, he tried to determine the flight speed of these bots. He estimated that a barely distinguishable brown blur, which he assumed to be a deer bot, rushed past him at a speed of about 400 yards per second. This translates to 818 miles per hour, considerably greater than the speed of sound, which is about 768 miles per hour. Townsend quoted a friend who wrote that "the idea of a fly overtaking a bullet is a painful mental pill to swallow." The friend was, of course, right; it is impossible. In a 1938 article in *Science*, the distinguished physicist and chemist Irving Langmuir debunked Townsend's oft-quoted estimate. Langmuir calculated that to maintain a speed of over 800 miles per hour, a fly would have to burn as fuel an impossible one and a half times its own body weight in food every second. He also pointed out that if a fly moving at that speed were to bump into a person, as sometimes happens, it would strike with the deadly force of a pistol bullet.

⚙ ⚙ ⚙   On a sunny summer afternoon, I saw a dense swarm of about a hundred tiny midges hovering about 4 inches above a small drop of dry white paint at the edge of the concrete slab that was my back porch. The swarm wavered slightly but remained fairly closely centered over the paint drop. When I swept through the swarm with my hand, the midges dispersed, but quickly reassembled over the same paint drop. Later I snapped up about 20 of them in a wide-mouthed pickle jar; they were so small that they would probably have gone through the mesh of my insect net. The next day I took them to my lab, immobilized them, and put them under the microscope. They were all males.

My impromptu observation confirmed what Frederick Knab, J. A. Downes, and many other entomologists had observed long before. Swarms of many insects assemble over "markers," such as the spot of white paint on my porch, that often lie on or near a horizontal border, such as the edge of my concrete slab. The border may be the margin of a lake, the edge of a field, or any other discernible line. The marker could be anything that contrasts with the background.

Knab saw mosquitoes by the hundreds hovering in swarms over

shocks of corn in a field; Downes saw swarms of midges hovering over wet spots or piles of cow dung that were darker than the sandy road on which they were located. To determine whether the swarms formed in response to moisture or odors rather than visually, Downes did a simple but convincing experiment. He placed three dark cloths on the road: one dry and free of odors, one that had been moistened with water, and one that had been impregnated with cow dung. Swarms assembled over all three cloths, leaving little doubt that the marker is perceived visually. A swarm can form and persist with little or no interaction between individuals. All that is necessary is that each individual that flies along the border and happens upon the marker stops and maintains its position by hovering so as to keep the marker in view directly below.

Downes and many others discovered that these swarms consist almost exclusively of males. For example, when Knab swept his net through a swarm of mosquitoes, he caught 897 males but only 4 females. Most likely a more or less steady stream of females arrive at the swarm, probably guided by the same borders and markers as the males. But when a female enters a swarm, she is immediately recognized, probably by the distinctive hum of her wings, and grasped by a male who quickly pulls her out of the swarm to copulate in nearby vegetation.

Swarms vary in size, from a few males hovering over a marker to dozens, hundreds, thousands, or even millions of individuals. Huge swarms of midges have been mistaken for smoke coming from a building that was thought to be on fire and even for great clouds of essential oils (secondary plant substances) arising from the tops of large trees.

There is no doubt that these swarms are mating assemblages, as has most recently been confirmed by M. Tokeshi and K. Reinhardt in a 1996 article detailing their studies of chironomid midges in a shallow lake in Northern Ireland. The larval midges live in the mud at the bottom of the lake as much as a thousand yards off shore. The adults emerge from the water en masse and fly to inland sites away from the shore, where they crowd together on foliage to rest as their bodies finish maturing. Once mature, the males form large mating swarms at the edge of the lake, using shoreside vegetation and light reflected from the water as markers. After mating, the females fly out over the lake to lay their eggs on the

water. Tokeshi and Reinhardt saw mating at the swarms but never at the resting sites or over the water.

⚙ ⚙ ⚙   While some insects have evolved special behaviors for attracting mates or setting up meeting places where widely scattered individuals gather, others can find a mate very easily, because large numbers of individuals of both sexes emerge as adults—sometimes by the millions—in the same place at the same time, probably because all of them are triggered to emerge by the same environmental signal, such as the increase of the temperature to some critical point. So it is with mayflies, the subject of the next chapter.

Myriads of Mayflies

*Adult mayflies clinging to cattail leaves at the edge
of the lake from which they emerged*

❀ ❀ ❀ Early on a summer evening in 1954, as I was driving along the main street of Janesville, Wisconsin, near the bridge that crossed the Rock River, I had to slow to a crawl and proceed very cautiously, because of an astonishingly dense swarm of mayflies. The broad-winged insects—almost an inch long, not counting their two even longer tails—fluttered everywhere and were so thick in the air that it was difficult to see more than a few feet ahead. The bodies of stunned and dead mayflies, already several inches deep on the street, made the pavement treacherously slick. Recently emerged from the river, probably late that afternoon and at dusk, the mayflies were being attracted to the brightly lit windows of several stores near the foot of the bridge. They beat themselves against the glass until they fell from exhaustion, and a deep pile of them was already accumulating on the sidewalk beneath the windows. When I drove by early the next morning there were even more dead mayflies. An endloader was scooping up huge piles of them that looked to be about 4 feet deep and loading them onto a dump truck.

A few species of mayflies, but by no means all of them, occur in comparably immense swarms. Early in the twentieth century, great swarms of mayflies, each swarm consisting of millions of individuals of only one species, emerged at intervals from the waters of the Great Lakes and the rivers of the Mississippi drainage system. The emergence of individuals within each swarm was closely synchronized. Beginning in the late afternoon and continuing into the night, all the millions of individuals in a swarm emerged within a few hours of each other. Several species did this—and many still do—but the ones most frequently seen on the Great Lakes and the rivers are two species of the genus *Hexagenia*. In 1944, F. Earle Lyman described dense swarms of *Hexagenia* that emerged at the western end of Lake Erie as being "so thick as to appear like clouds of smoke, weaving up and down over the tree-tops." Not long thereafter, as you will read below, water pollution extirpated these insects from much of their former habitat, although in many areas they are now making a dramatic comeback as the environment is cleaned up.

Mayflies have persisted since geologically ancient times. They belong to the most ancient surviving lineage of winged insects. Their fossils are found in deposits that were laid down over 300 million years ago during the Carboniferous period; the very individuals that were fossilized probably flew near or among the tree ferns and other primitive plants of the great coal-forming swamps of that time. Today there are somewhat over 2,000 extant species worldwide and 625 in America north of Mexico.

As is to be expected, the insects of this ancient line have some primitive characteristics. They are the only insects that have two adult stages. After the first adult stage molts its nymphal skin on the water, it flies to the shore, where it rests until it molts to the second and last adult stage, in which reproduction occurs. In common only with the dragonflies and damselflies, which together form a group almost as ancient as the mayflies, they have not evolved the hinges and muscles needed to fold their wings down flat upon the body, as do most other insects. They can hold them only out to the sides or together straight up over the back. This is no small matter. The ability to fold their wings out of the way made it possible for more advanced insects to evolve into hundreds of thousands of species with a multitude of life styles that require them to get into tight places: to crawl under rocks, to tunnel beneath the leaf litter of the forest, or to burrow into the soil or into plant stems, leaves, or fruits.

❁ ❁ ❁   The mayflies belong to the order Ephemeroptera. Loosely translated from its Greek roots, it means "winged ones that live for only a day," an apt description of almost all adult mayflies. Although the adults of a few species may survive for a week or more, most live for only a day or two and a few for less than an hour. All mayflies spend almost their entire lives, in some species 364 days of the year, as aquatic nymphs.

As John Brittain explained, mayfly nymphs have paired and usually platelike gills on the abdomen that absorb dissolved oxygen from the water and that in some species double as paddles for swimming. The nymphs molt many times—another primitive characteristic—in some species about 20 times and in a few as many as 50 times. Each species has its own preferred habitat. Mayflies are found in swift-flowing streams, slow-moving rivers, and ponds and lakes of all types. Some cling to rocks so as not to be swept away by a swift current, some crawl about in aquatic vegetation, and many burrow in the mud or silt on the bottom of a slow-moving river or a lake. Although a few are predators that eat other aquatic insects, the great majority are not predaceous but feed on plankton (mostly tiny algae) and organic detritus that they filter from the water, while others scrape algae and other tiny organisms from the surface of aquatic plants or other objects in the water. As nymphs, the two species of *Hexagenia* that emerge from lakes and rivers in huge swarms live in the bottom muck in shallow burrows with an opening at each end, filtering tiny bits of food from the water—probably from a current that their waving gills force through their U-shaped burrows.

When *Hexagenia* mayfly nymphs are fully grown, they leave their burrows to swim up toward the light and break through the surface film of the water. The nymph's skin immediately splits, and the first-stage adult bursts out, stands on the surface film briefly until its wings harden, and then flies to shore, where it rests on vegetation, buildings, or other surfaces—doing nothing until it molts to the second adult stage the next day.

At dusk, second-stage adult males gather near trees or other vegetation on the shore to congregate in airborne mating swarms. In 1976 George Edmunds, Jr., and two coauthors described a mating swarm of male *Hexagenia* on the shore of a lake in Michigan: Most of the males flew at the level of the treetops, but some flew even higher. Within about 10 minutes, most of the nearby males had joined the swarm in its bobbing flight, each male "flying rapidly upward for several feet and then drifting downward to rise again." Females soon began to fly into the swarm and were quickly seized by males. As a pair copulated, they kept on flying but slowly dropped downward, and after no more than 30 seconds they separated while still in the air. The males usually returned to the swarm, which continued to fly until just after dark. Then all the males landed and remained still until they died soon thereafter.

Immediately after separating from their mates, the females flew directly out over the lake to get on with the all-important and all-consum-

ing business of laying eggs. They usually dropped to the water's surface and sprawled there with their wings widely spread out to the sides. Then they raised the tip of the abdomen upward and extruded their eggs, on the average about 4,000 of them. But some individuals were snapped up by fish before they could finish or even start laying. Surviving females flew off after they finished laying, but soon died. The egg-laying period was short. Within half an hour after the first mated females arrived over the lake, all or almost all of those that had mated had finished laying their eggs.

⚙ ⚙ ⚙   Adults of both stages are totally defenseless, and are freely eaten by the various predators in their vicinity. As James Needham and his coauthors wrote, "We have seen robins at the waterside with bulging crops and half a dozen mayfly tails hanging out of the angles of their beaks. Other birds search the foliage for them; others harvest them when they swarm. Spiders lay snares that capture them by the thousands . . . Dragonflies of many kinds hover about the swarms, capturing them at will, as hawks follow coveys of pigeons and quail. Bullfrogs sit by the waterside, alert to catch them whenever they fly past, and the return of the females to the water for egg-laying marks the renewal of attack by carnivorous fishes. Trout, bass, pickerel, and many other fishes snap them up at the surface, or even spring into the air to seize them while in flight." Many people who fish with artificial flies are well acquainted with mayflies. When a "hatch" of mayflies appears on their favorite trout stream, they use as artificial bait "flies" made of feathers and hair that resemble mayflies.

Insect eaters take full advantage of a mass mayfly emergence to gorge themselves, but even though the mayflies are virtually defenseless, barely capable of dodging the peck of a beak or escaping by flying away, many of them survive to reproduce. As many biologists have written, and as I wrote in earlier chapters, emerging in swarms that are closely synchronized in time and space increases the individual's probability of finding a mate. And of course it also decreases the probability that an individual will become a meal for a predator. As is the case with monarchs, locusts, and periodical cicadas, there are many more of them than the predators can eat. The predators are fully satiated by the time they have eaten a small part of the swarm.

In studies of an unusual mayfly, *Dolania americana,* in Upper Three Runs River in South Carolina, Bernard Sweeney and Robin Vannote found that the larger the number of mayflies in an emergence the smaller the percentage of them that is eaten by predators, thus confirming the predator satiation hypothesis. *Dolania* nymphs burrow in the sandy bottoms of streams and are predators that eat mainly the aquatic larvae of midges. Some adults emerge each day during the first 2 weeks of June. Between an hour and half an hour before sunrise, fully grown nymphs swim to the surface of the water and molt to the adult stage in less than 20 seconds as they float downstream. After molting to the second adult stage, males fly back and forth over the river as they search for females. Female *Dolania* are unusual among mayflies in that they reproduce during the first adult stage and never molt to the second adult stage; they are usually mated within a few seconds after they fly from the surface of the water and after that immediately lay their eggs. Females seldom survive for more than half an hour, although males may live for as long as an hour.

On some mornings during the emergence period, many *Dolania* are present, but on other mornings they are much less numerous, thus providing a natural experiment for determining the effect of the size of a swarm on the survival of the mayflies. Since these smaller swarms contain at most a few hundred individuals, Sweeney and Vannote were able to keep tabs on the number that emerged and the number that survived. Downstream from the swarming site, they stretched a seine across the river to catch the floating molted skins of the nymphs and the floating dead bodies of adults, virtually all of which fall to the surface of the water. By subtracting the number of adult bodies, which represented the survivors, from the number of nymphal skins, which represented the total number of mayflies that emerged, they obtained the number of individuals that had disappeared, presumably eaten by predators. On mornings when the swarms were quite small, 30 or less, from 80 to 90 percent of the adult mayflies of both sexes were missing, presumably eaten by such aerial predators as bats, nighthawks, swallows, and dragonflies. When swarms were larger, from 100 to 250 indi-

viduals, the percentage eaten was much lower, only from about 20 to 30 percent for females but as much as 50 percent for males. Males are more exposed to aerial predators because they spend much more time in the air than do females.

❀ ❀ ❀   The mayflies that once emerged in great swarms from the western end of Lake Erie were totally wiped out by water pollution, mainly by municipal wastes from Detroit, Toledo, and smaller cities. As Kenneth Wood reported, samples taken from the lake bottom from 1929 to 1952 showed that the population of *Hexagenia* nymphs in that part of the lake had been reduced by 90 percent by 1952. By 1961, virtually all of them were gone from the western end of the lake.

In that year, John Carr and Jarl Hiltunen compared the level of pollution and the population of mayfly nymphs that was then present in a 387-square-mile area at the extreme western end of Lake Erie with the level of pollution and the population that had been found in the same area by another researcher in 1930. The results were disheartening, to say the least. Where bottom samples had shown an average of over 116 large mayfly nymphs per square yard in 1930—as many as 530 per square yard in the open lake—in 1961 there was less than one; in fact, only 10 nymphs were found in all of the many samples taken in the area that year.

In 1930, there was more pollution than there would have been if municipal wastes had been properly managed. Nevertheless, over 66 percent of the 387 square miles had not yet been polluted. Nineteen percent of the area was lightly polluted, and the 14 percent that was moderately to heavily polluted was confined to the mouths of three rivers: the Detroit, which flows past Detroit and Windsor, Ontario, the Raisin, which flows through Monroe, Michigan, and the Maumee, which flows through Toledo, Ohio. But by 1961, the entire area was polluted. Twenty-six percent, mainly open lake well away from the shore, was only lightly polluted. All the rest was moderately to heavily polluted. The areas of heavy pollution, 23 percent, were at or near the three river mouths that had been moderately to heavily polluted in 1930 but were much larger and far more heavily polluted in 1961.

The pollutants, mainly sewage and runoff from heavily fertilized farm fields, were heavily laden with the nutrients nitrogen and phosphate, thus causing the lake to become *eutrophic,* a word that can be translated into plain English as "overfertilized," as Paul and Anne Ehrlich put it. As eutrophication proceeded, the overabundance of nitrogen and phosphate completely disrupted the ecology of the lake. As the Ehrlichs wrote, these nutrients made possible the growth of immense populations of algae. When these algae died, their subsequent decomposition depleted the water of oxygen and killed off animals, such as the mayfly nymphs, that have high oxygen requirements. The disappearance of the mayflies and other aquatic insects, important components of the diets of fish, were an important factor in the catastrophic decrease in the fish population of Lake Erie.

Moreover, the decrease in populations of aquatic insects, mainly mayflies, further accelerated the rate of eutrophication. These insects remove nitrogen, phosphate, and other nutrients from the lake and incorporate them in their bodies. Many of the adult mayflies in a swarm, probably including almost all the males, die on shore and thereby transfer nutrients from the lake to the land. When mayflies were abundant, millions of pounds of their bodies fell to the land, and as long as the level of pollutants entering the lake was low, this removal of nutrients had been enough to keep the lake from becoming overfertilized.

While mayfly nymphs virtually disappeared from 1930 to 1961, there were immense increases in the populations of bottom-dwelling creatures that thrive in polluted areas where oxygen levels are low. All of these creatures are reliable indicators of pollution. Of the animals found in an average bottom sample in 1961, 84 percent were sludge worms, so named because they live in sewage sludge; of the remainder, 8 percent were fingernail clams and 5 percent were the larvae of midges, both tolerant of pollution. From 1930 to 1961, the average population of sludge worms increased ninefold. Near the mouth of the Detroit River, which had already been moderately polluted in 1930, the average number of sludge worms found in that year was about 217 per square yard, but had increased to over 7,500 by 1961, with one sampling station having an astronomical 32,740 per square yard.

Lake Erie was only one of many polluted bodies of water in the

United States in the 1960s. For 200 years of American history, sewage and industrial wastes had been heedlessly dumped in many of our rivers and lakes. In the *short run,* that is the cheapest way to get rid of them. Towns and businesses were interested in short-run savings and profits and gave no thought to long-run consequences, including economic losses that ultimately more than wipe out the short-run gain from dumping wastes in rivers and lakes. By the 1960s, some rivers were virtually open sewers and were best known for their stink. Many were not safe to swim in. Salmon, striped bass, and shad no longer ran up most of the rivers of the east coast. The Cuyahoga River, which runs through Cleveland to Lake Erie, was so fouled with industrial wastes that it burst into flames in June of 1969.

This almost unimaginable occurrence, a river on fire, was, as Paul Schneider put it in an article in *Audubon,* "a major impetus to the passage of the Clean Water Act" by Congress. Since then many rivers have been cleaned up. Pollution has been largely eliminated from Lake Erie. Even the once unutterably foul Cuyahoga is now an asset to Cleveland and, as Schneider wrote, a part of it "is now lined with restaurants and pleasure boat slips." And, Schneider continued, Lake Erie, "widely perceived to be 'dead' in the late 1960s, is today touted as the 'walleye capital of the world.' Bacteria counts and algal blooms dropped more than 90 percent between 1968 and 1991; now, like the other Great Lakes, Erie is officially safe for swimmers along 96 percent of its shoreline, and lake-based tourism contributes more than $8 billion a year to the Ohio economy." But much remains to be done to clean up the nation's lakes and rivers. Eutrophication is not the only problem. The Great Lakes are still so polluted with insecticide residues and PCBs (polychlorinated biphenyls) that states along their shores warn against eating salmon from the lakes. Recent attempts by Congress to gut the Clean Water Act and the Environmental Protection Agency warn us to be constantly vigilant.

❀ ❀ ❀   In a recent conversation, Lynda Corkum of the University of Windsor in Windsor, Ontario, an expert on the mayflies of Lake Erie, told me that the mayflies are back, although their distribution is somewhat patchy; in some places there are few and in other places there are many. Using a sampling device known as a Ponar grab, she scooped samples from the bottom muck in 1997, and found that *Hexagenia* nymphs are once again plentiful: about 1,180 of all sizes per square yard off Toledo, 725 per square yard off the mouth of the Raisin River, and about 195 per square yard near the mouth of the Detroit River. Huge swarms of adults are now fairly frequent at the western end of the lake. In 1996 adult mayflies were so astronomically abundant in Port Clinton, Ohio, that snowplows had to be used to clear them from the streets. In 1998, they were so numerous at Colchester, Ontario, that a snow-removal machine was used to clear them from the beaches.

Corkum attributes the recovery of the mayfly population not only to human efforts to eliminate pollution, but also to the zebra mussels that were accidentally introduced into the Great Lakes, probably from the Black and Caspian Seas, in the ballast water in ships. They were first seen in Lake St. Claire, the source of the Detroit River, in 1988 but have since spread widely throughout the Great Lakes as well as to many other bodies of water. They are serious pests because great clumps of them grow in and plug the water-intake pipes of power plants and other water-using installations. Zebra mussels are filter feeders that strain bits of detritus and plankton, tiny algae and other organisms, from the water and deposit them as feces on the bottom, thus helping to diminish eutrophication and to make the water of Lake Erie very clear. So far this has been a help to the mayflies, but as Patrice Charlebois of the Illinois Natural History Survey pointed out to me, the pestifrous mussels are also competition for the mayflies and other filter feeders in the lake.

❀ ❀ ❀   Great swarms of *Hexagenia* mayflies also occur in parts of the upper Mississippi River, especially in the 670-mile stretch from St. Louis, Missouri, north to Minneapolis, Minnesota. This segment of the river is divided into impoundments by 29 navigational locks and dams built by the U.S. Army Corps of Engineers, beginning in 1917. The silt that inevitably settled in these impoundments is a very favorable habitat for burrowing mayflies such as *Hexagenia* and made possible large increases in their populations. As William Mason and two coworkers reported, surveys made from 1957 to 1969 and in 1976 showed that most of the navigation impoundments upstream from Minneapolis and St.

Paul, the Twin Cities, supported large populations of *Hexagenia* may-flies.

But they "were conspicuously rare" in the 45 miles downstream from the Twin Cities and even farther downstream in Lake Pepin, a natural impoundment that trapped the pollution from the Twin Cities. The sewage and other wastes from these cities, which have a population close to 2 million, caused severe eutrophication and all but eliminated mayflies from Lake Pepin and the aforementioned segment of river below the Twin Cities—just as pollution had eliminated them from Lake Erie.

After the passage of the Clean Water Act, the Twin Cities improved their sewage-treatment facilities and eliminated much of the pollution that they had been discharging into the Mississippi. As Calvin Fremling and D. Kent Johnson reported, the downstream mayfly populations recovered. In 1986, 21 *Hexagenia* swarms emerged from the river near and downstream from the Twin Cities. The insects were numerous enough to be a nuisance in St. Paul. The largest emergence occurred on the night of June 23, 1987, when "snowplows were needed to clear the Interstate Highway 494 bridge in St. Paul."

Although mayflies can be a nuisance, people are beginning to understand that they are also abundant food for fish, and are an indication of water purity and the ecological health of a body of water. The disappearance of or large decreases in the populations of certain aquatic insects, notably mayflies and stoneflies, is a reliable indication of pollution. Chemical analyses of the water and the bottom sediments can tell us which pollutants are present and in what quantity. But they do not measure the biological effects of pollution. For instance, they cannot tell us how different pollutants and other stress factors interact with each other to affect living creatures. The ability of mayfly or stonefly nymphs to survive and grow is a measure of the total combined effect of all the pollution and other stresses to which the nymphs are exposed.

✸ ✸ ✸   Like mayflies, locusts, and some other insects, periodical cicadas emerge synchronously as adults by the millions, thus greatly facilitating the coming together of the sexes. But as you will read in the next chapter, an emergence of periodical cicadas is considerably more complex than an emergence of mayflies. A cicada emergence includes three similar but distinct species that do not cross-mate. The members of each of these three species locate each other by forming separate singing aggregations within the larger mixture of the three species. Their songs are all different, as can be distinguished even by the human ear.

Swarms of Cicadas

*A female periodical cicada piercing a twig as she gouges
out a hollow in which to lay her eggs*

⚙ ⚙ ⚙   "A great observer, who hath lived long in *New England,* did upon occasion, relate to a Friend of his in *London,* where he lately was, That some few Years since there was such a swarm of a certain sort of Insects in that *English* Colony, that for the space of 200 Miles they poyson'd and destroyed all the Trees of that country; there being found innumerable little holes in the ground, out of which those Insects broke forth in the form of *Maggots,* which turned into *Flyes* that had a kind of taile or sting, which they struck into the Tree, and thereby envenomed and killed it." So wrote Henry Oldenburg in the first published account of periodical cicadas, which appeared in 1666 in the very first volume of the *Philosophical Transactions of the Royal Society of London,* a scholarly journal that is still published today. Much of what Oldenburg wrote is inaccurate, but there is no doubt that it refers to periodical cicadas. The "Maggot" is, of course, the nymph, the immature stage, which comes out of the soil through one of those "innumerable little holes in the ground," crawls a short distance up the trunk of a tree, and sheds its skin to become the "Flye," the winged adult cicada. The "taile or sting" is the ovipositor, which the female uses to puncture the twig of a tree and create a small cavity in which to lay a batch of eggs. The poisoning and destruction of the tree is a gross exaggeration. Periodical cicadas do not kill or seriously injure large trees, although they may slow the growth of small saplings. But even on large trees, there is likely to be some very apparent but actually insignificant damage that results from the hundreds of egg punctures a tree may sustain. Many twigs break at the site of a puncture and die but for some time continue to hang down like flags, giving the tree a half-dead look when the leaves on these partially severed twigs turn brown. I am surprised that Oldenburg's informant did not mention the deafening noise made by the cicadas as thousands of males sing in chorus to attract females. Oldenburg compared the cicadas to locusts, which, as you know, are actually grasshoppers, and sometimes occur in huge, damaging swarms in Europe. This may have started the incorrect use of the word "locusts," which persists to this day, to refer to periodical cicadas.

There are much earlier accounts of cicadas in ancient Old World writings—not of periodical cicadas, which are found only in the New World, specifically the eastern United States—but rather of solitary and nonperiodical species. As far as I know, cicadas are not mentioned in the Bible, but as J. G. Myers has noted, they are often mentioned in the writings of the ancient Greeks and Romans, beginning with Homer's *Iliad* in the eighth century B.C.E. Aristotle, the most perspicacious of the ancient observers of nature, wrote a scientific account of cicadas that is remarkably accurate, although it does include a few flights of fancy. Most other ancient writers did not view cicadas from a scientific perspective. For example, they were well aware of the cicada's song but focused on its musical merit rather than its biological significance. As Myers said, "It is a fact that ever excites the wonder of the commentators, that most of the Greek classical references to cicada song are highly laudatory. Yet the Latin references, dealing apparently with the same species, characterize it as raucous and disagreeable." The Greeks even kept cicadas in cages for the pleasure of hearing them sing, and Aristophanes took a favorable view of the singing of cicadas in *The Birds:*

> But in flowery meads I dwell,
> Lingering oft in leafy dell,
> When the inspired cicada's gladness,
> Swelling into sunny madness,
> Filleth all the fervid noon
> With its shrill and ceaseless tune.

Myers went on to note that "some [modern] writers have even gone so far as to use Greek praise of cicada song as a foundation for impugning Greek taste in music altogether."

Of the 180 species of cicadas that occur in North America, (1,940 species worldwide) the adults of two groups of species are conspicuous and familiar to many people: the spectacularly abundant and gregarious periodical cicadas and the dog-day cicadas, the latter loners that are not gregarious and never occur in overwhelming abundance. Their life cycles last for several years, but some are present every year because their generations are staggered. Periodical cicadas are so called because the adults appear periodically, emerging from the soil in close synchrony exactly once every 13 or 17 years. The dog-day cicadas were so named because the adults appear during the "dog days," the hottest days of the summer. Cicada nymphs of either group are seldom seen because they burrow in the soil to suck sap from the roots of woody plants. They emerge from the ground and molt to the adult stage in the dark, but their cast skins, looking like pale beige insect ghosts, can be found the next day clinging to the trunks of trees near ground level.

Male cicadas sing to attract females, which are mute but can hear, for the purpose of mating. Their sound-producing organs are said to be more complex than those of any other animal on earth, certainly more so than the larynx of a human or the syrinx of a bird. They are very different from the sound-producing organs of all other insects. Other insects produce sound by stridulating, by rubbing one part of the body against another, as a grasshopper rubs a leg against a wing or as crickets and katydids rub two wings together. The sound-producing organs of cicadas are in two large cavities, one on either side of the base of the abdomen. Each is covered by a flap and contains a convex timbal, or drumhead, to which is attached a powerful muscle. The sound is produced when the flap is lifted and the timbal is repeatedly drawn inward and released. Sir Vincent Wigglesworth of Cambridge University, one of the greatest insect physiologists of all time, compared this action to pressing down and releasing the bulging lid of a tin can. Pushing the lid down produces a click, and releasing the lid produces another click. Cicadas make their high-pitched sound by producing about 390 clicks per second. While people perceive the sounds produced by birds as sweet and melodious and are charmed by the chirp of a cricket, most people find the sounds produced by cicadas shrill and strident. But to a female cicada the song of a male is apparently enticing and seductive.

Male periodical cicadas form singing choruses that produce what is to our ears a deafening cacophony. Dog-day cicadas sing loudly but do not join together in choruses. On hot afternoons in July and August, lone dog-day cicadas, much more often heard than seen, sing their droning songs high in trees, including shade trees along suburban streets. Although they do not form organized choruses, many may sing near each other at the same time and make a considerable noise.

Dog-day cicadas live from 2 to perhaps as many as 9 years. But gener-

ally speaking, some individuals of all species appear every year, because their generations overlap; their life cycles are staggered rather than periodical. Dog-day cicadas are the natural prey of cicada killer wasps, which begin to appear in early July, in time to hunt the dog-day cicadas but too late to exploit periodical cicadas. These huge wasps—their body length approaches 2 inches—paralyze but do not kill dog-day cicadas with a sting and then carry them to their burrows in the soil, where they will eventually serve as food for the wasps' larvae. On a hot day in August, I saw a female cicada killer leave her burrow and fly to a nearby clump of trees, presumably in search of a dog-day cicada. A lone male cicada droned in the trees, but not long after the cicada killer had flown off, his drone suddenly ended in a squawk. The wasp soon returned to her burrow carrying a male cicada. The evidence is circumstantial, but there isn't much doubt that it was the cicada killer which brought that song to a sudden end. But cicada killers do not locate their prey by sound; they actually catch more mute females than noisy males. They presumably find their prey visually or by odor.

❀ ❀ ❀   Periodical cicadas are remarkable because of their long life spans and the periodic appearance of immense numbers of them once every 13 or 17 years. Although other insects may live as long or even longer—a queen ant, for example, survived in a laboratory colony in Germany for 29 years—periodical cicadas have by far the longest growing stage of any known insect. Like dog-day cicadas, the nymphs live in the soil feeding on sap that they suck from the roots of trees or other plants. When the period of growth is over, they emerge from the soil en masse as a tightly synchronized group, molt to the adult stage, and go about the business of reproduction. At the latitude of central Illinois, the emergence begins in mid-May and adults will be present for about the next 4 weeks and then will not be seen again for another 13 or 17 years.

Entomologists define a "brood" of periodical cicadas as all the populations that emerge in the same year and are more or less geographically contiguous. Generally speaking, the broods that emerge once every 13 years occur mainly in the lower midwest and the south, while those that appear once every 17 years are concentrated farther to the north and east. But there is some overlap. For example, 13-year and 17-year cicadas occur near each other in Missouri, Kentucky, Illinois, Indiana, and a few other states. However, as Kathy Williams and Chris Simon have pointed out, a given pair of 13- and 17-year broods will emerge in the same year only once every 221 years. Thus two broods that emerged together in 1897 will not emerge together again until 2118.

In 1898, Charles Marlatt recognized 30 different broods of periodical cicadas, 17 17-year broods that he designated I to XVII and 13 13-year broods that he designated XVIII to XXX. According to Williams and Simon, only 15 of these broods are currently extant, 12 17-year broods and 3 13-year broods. The missing broods could be accounted for in several ways. Some of those recognized by Marlatt may have been only rumored to exist, and others, then represented by only a few individuals, may already have been on the road to extinction. Some of the broods mapped by Marlatt are known to have since gone extinct, often because of human activities such as the clearance of forests for agriculture.

During the few weeks before they metamorphose to the adult stage, the fully grown periodical cicada nymphs leave the roots from which they have been feeding and extend their tunnels up to the surface of the soil. In 1979 in Connecticut, Chris Maier discovered that all nymphs had moved to positions just below the soil surface 19 days before the first of them actually left the soil to molt to the adult stage. Before leaving the soil, some of them build around their emergence holes tubelike turrets, or "chimneys," of mud pellets that resemble the much larger turrets built by crayfish on low-lying wet soil. The nymphs wait at the mouth of the burrow for the environmental signal, possibly the increase of the soil temperature to some requisite critical level, that tells them to leave the soil forever. When a nymph leaves its tunnel, usually just before or after sunset, it crawls a short way up a tree trunk or some other support and molts its hard skin, becoming a pale and as yet unhardened adult whose new skin is so soft and delicate that the poke of a finger can split it and doom the insect to death from loss of blood. The skin of the newly molted adult soon begins to harden and darken, but the process will not be completed for some time. The ovipositor of the female and the sound-producing organs of the male, for example, take several days to become completely hardened.

Most of the nymphs in a population become adults within 7 to 10 days of each other, resulting in the accumulation of almost unbeliev-

ably immense populations of winged adult cicadas. According to Henry Dybas and D. Dwight Davis, the density of periodical cicadas in a 17-year brood (XIII) that emerged in the Chicago area in 1956 varied from an average of about 133,000 individuals per acre in an upland forest characterized by oaks and hickories to about 1,500,000 per acre in a low-land forest characterized by elms, ashes, and hackberries. The average total weight of cicadas in the lowland forest was about 1.2 tons per acre, and the maximum was about 1.6 tons per acre. This amounts to almost a billion cicadas and a total weight of 533 tons per square mile of lowland forest. And the emergence extended over several hundred square miles.

Such an abundance of protein-rich food is, of course, a tremendous windfall for the insect-eating predators that live in or near the area of an emergence. It is a gastronomic boon with no strings attached. The cicadas are essentially defenseless, an easy meal for any insect eater that comes along. They are not camouflaged, they do not hide, and they are easy to catch because they are generally not wary. They cannot sting and they are neither toxic nor nasty tasting, as are monarchs and many other insects. As a matter of fact, some people, and presumably other creatures too, consider cicadas to be delicious and eminently edible.

By contrast, the nonperiodical cicadas, which are solitary and do not occur in aggregations, are extremely wary and very difficult to catch. Some have very specialized defenses. Hugh Cott noted that one found in Malaya is brilliantly colored in black and scarlet. This could be a warning of its inedibility, but could also be a case of Batesian mimicry, an edible insect bluffing by resembling some other toxic insect. Cott also describes the presumably protective "flash colors" of a forest-dwelling nonperiodical cicada of the Amazon Valley:

Most Cicadas have the wings quite transparent and colourless, but in the present species the innermost, or proximal [basal], portion of each hind-wing is decorated with a vivid splash of vermilion, which extends nearly half-way along the wing. This colour would, of course, be perfectly visible through the transparent fore-wings in the resting insect, were it not for the fact that a patch of about equal area on these wings is pigmented with opqaue olive-green, so that when the wings are folded the cryptic green area closes like a shutter so as just to hide the red areas on the wings beneath.

The brilliant "flash color" is revealed when the insect takes flight at the approach of a predator, perhaps startling the predator and thus giving the insect precious seconds in which to escape. The red is again concealed when the insect alights, perhaps misleading a pursuing predator that has a "search image" for brilliant red.

⚙ ⚙ ⚙   When brood XIII reappeared in the Chicago area right on schedule in 1990 a few people panicked and tried to stem the hordes by applying insecticide, an environmentally harmful and futile endeavor. The cicadas are harmless and there are too many of them to kill. But most people, forewarned by the media, greeted the cicadas with equanimity. Many just sat back and enjoyed the spectacle, and some took advantage of this abundance to feast on cicadas. With their legs and wings removed and their bodies suitably seasoned, cicadas are eaten raw, deep-fried, sauteed, stir-fried, roasted, or boiled. Most people who have tried them like the flavor. Roasted cicadas, for example, are said to taste "meaty and delicious." Bars held cicada-eating contests, and recipes for preparing cicadas appeared in newspapers and magazines. *Time* magazine recommended that they be dipped in batter, fried until golden brown, and served with cocktail sauce or sour cream. The anonymous author of an article entitled "When Chicago Braced for the Onslaught of the 17-Year Cicada," taking note of the many cicada recipes that had appeared in print, observed that "Americans consider cicadas, unlike most insects, as almost respectable food. Maybe, someday, a predicted periodical cicada emergence will automatically call for a community festival. The cicadas would provide not only the food, but also the music."

Wild creatures—birds, mammals, and even reptiles—feast on periodical cicadas whenever the opportunity arises, and presumably find them as delicious as do people. Birds are the most prevalent consumers of these insects, including, among others, ducks, terns, gulls, Mississippi kites, cuckoos, red-headed woodpeckers, blue jays, titmice, wood thrushes, robins, vireos, starlings, grackles, red-winged blackbirds, and house sparrows. Among the mammals that have been observed to eat periodical cicadas are racoons, domestic dogs and cats, shrews, squirrels, voles, and mice. Even snakes and turtles take advantage of this

plethora of edible insects. Not only do all of these creatures eat cicadas, but they, notably the birds, shamelessly gorge themselves on these insects. Thomas Moore of the University of Michigan, an authority on periodical cicadas, has watched birds bingeing on these insects, cramming down so many of them that their bodies become noticeably swollen and sometimes even grossly distended. Tom told me that he saw a starling that was so bloated it could not fly. He was able to run it down and could easily have caught it if he had wanted to.

✿ ✿ ✿   It may be that the sounds produced by cicadas give them some limited protection against predators. Perhaps the loud squawk made by a male when it is attacked will startle and maybe even discourage an attacking bird. The overall uproar of an aggregation of these insects may have a greater effect. The massed sound of an emergence of periodical cicadas is nearly deafening to humans. It is at a level of at least 80 decibels, 10 times as noisy as traffic on a freeway and 10 million times as noisy as rustling leaves. James Simmons and his research group speculated that this noise repels birds in two ways: First, it is so intense that is is painful to their ears. Second, it is so pervasive that it makes it difficult for birds to hear each other, thus interfering with their most important means of communication, their calls and songs.

Observations made by Simmons and his colleagues support these ideas. When cicadas are making a lot of noise, there are few birds in their vicinity—often none in the center of a congregation of noisy cicadas and usually only a few around its fringes. But "when the sun was obscured by clouds, and the sound production was less, there were more birds in the area and they stayed longer." Nevertheless, some birds fed on cicadas even when the latter were at their noisiest. "Certain birds—especially grackles—were observed to fly into the area, snatch a cicada, and at once fly away to eat it." But judging by the many periodical cicadas that are eaten by birds—as indicated by the many discarded wings and partly eaten bodies that litter the ground—this presumed defense is at best only partially effective. After all, there are times when the cicadas are less noisy than usual, and there are birds that attack them despite the noise.

How, then, do periodical cicadas survive, largely defenseless, delicious, and beset on all sides by animals that relish them and eat as many of them as they can cram down? Some writers have put this in another way, asking how it is that these insects can afford to be "predator-fool-hardy." The answer, as you well know, is that there is safety in numbers. There are so many cicadas that the insect-eating animals in the vicinity, no matter how gluttonous, can eat no more than a relatively small fraction of them. In other words, the cicadas are so abundant that they, like locusts, sate the appetites of their predators. As Robert and Janice Matthews observed, "to escape predation, it is better to live in an area where predators have full stomachs rather than where their stomachs are empty." The combined force of avian, mammalian, and reptilian predators will eat hundreds of thousands or even millions of cicadas, but many other cicadas, usually millions of them, will survive to reproduce their kind.

The hypothesis that populations of adult periodical cicadas are large enough to saturate the appetite of predators is strongly supported by two sorts of observations that have been made under natural conditions. First, the number of cicadas taken per day by predators, generally only a small fraction of those present, tends to remain more or less constant as the emergence progresses and the population of adult cicadas increases. This is so because the number of predators present stays more or less the same, and they can eat only so much and will not eat more no matter how many more cicadas become available. This is, of course, not a hard and fast rule. The number of cicadas eaten will increase if more predators, such as flocks of grackles or red-winged blackbirds, invade the area, and—later in the season—as young birds in the nest demand an ever increasing quantity of food. The *proportion* of the cicada population that is consumed by predators will, of course, be higher than otherwise when cicadas are the least abundant, early in the season before most of them have emerged from the soil and again late in the season when many of them have already died of old age. Observations of a 1985 periodical cicada emergence in Arkansas by Kathy Williams and two co-researchers showed that most periodical cicadas survive the onslaught of predators. They estimated that predators killed only from 15 percent to 40 percent of the cicada population, mostly old and declining individuals that had already mated and laid eggs.

The second supporting observation is that very small populations, apparently too small to saturate the appetites of predators, suffer very heavy predation and are sometimes totally annihilated by predators, mostly birds. In May of 1961, when an emergence of periodical cicadas was just beginning in Roockingham County, Virginia, Richard Alexander and Thomas Moore observed that a flock of unidentified blackbirds came to feed on the few cicadas that had already emerged. These early emergers were all destroyed; "not a single living adult cicada could be found or heard, although wings and mutilated individuals could be picked up around the bases of large trees bearing nymphal skins on their trunks." However, most of the cicadas that emerged at a later time in larger numbers survived.

Researchers who transplanted eggs to new habitats found that when the adult cicadas emerged 17 years later, only a few thousand per acre, they were all eaten by birds and other predators before they could lay any eggs. When Alexander and Moore moved about a thousand adults to a cicada-free habitat, the cicadas were almost immediately attacked by birds and "within three days there were no cicadas to be heard or seen in the entire area."

❁ ❁ ❁ Do these insects benefit from their long-term periodicity? The answer is that it certainly appears that they do. A predator or a parasite could theoretically prosper and maintain an unusually large population by specializing on adult periodical cicadas, taking advantage of their abundance by synchronizing its life cycle with theirs by remaining dormant during the 13- or 17-year interval between emergences, a time when no comparably abundant food supply is likely to be present. But so far no animal of any sort has managed to exploit periodical cicadas by solving the problem of remaining dormant without feeding for such a long period of time. Only one organism, the parasitic fungus *Massospora cicadina*, has evolved the ability to synchronize its "feeding period" with the cyclical appearance of adult periodical cicadas. Cicada nymphs are infected by virulent spores of this fungus as they emerge from the soil to metamorphose to the adult stage. Before the parasitic fungus kills the adult cicada, it produces resting spores that fall to the ground and lie dormant for the 13 or 17 years until the next generation of cicadas emerges from the soil.

It may not be coincidental that the numbers designating the years between succeeding generations of periodical cicadas, 13 or 17, are both
odd numbers. They also happen to be prime numbers, evenly divisible
only by one. Monte Lloyd and Henry Dybas have suggested that when
periodicity was evolving, a generation emerged over several years with
a sharp peak in the middle year. If, for example, emergences continued
for a span of 8 years, and their peaks were separated by 16 years, a parasite or predator could have exploited a hypothetical "protoperiodical"
cicada by coinciding with the first individuals of an emergence, remaining dormant for a plausible 8 years, not becoming active again until the
last cicadas of that emergence appeared, and then remaining dormant
for another 8 years until the first cicadas of the next generation appeared, and so on. But if the generation peaks became separated by 17
rather than 16 years, the cicadas would—even this early in the evolutionary game—have escaped the parasite or predator, which could not
have stayed in synchrony with them because the gap between the last
emergers of one cicada generation and the next would have been longer
than the gap between the early and late emergers of a single generation.
A similar argument can be made for 13-year broods.

✖ ✖ ✖   Massive aggregations of periodical cicadas form because
their members are synchronized in growth and development and because all are triggered to emerge at about the same time by the same environmental cue. Their synchronization is tightened by natural selection, because individuals that emerge before or after the main body of
the crowd are more likely to be eaten by predators and, even if they survive, are less likely to find a mate and leave behind offspring.

Although the overall aggregation is essentially unstructured, the situation is actually more complicated than it at first seems to be. For one
thing, most broods include three different and distinct species that are
reproductively isolated from each other—that do not hybridize even
though they emerge in unison and live in close proximity. The same
three species occur in 13- and 17-year broods. The other complication is
that smaller, structured, cohesive groups of singing males form within
the overall massive aggregation. The males of each of the three species
attract mates by forming large singing choruses that come together be-

cause the males as well as females are attracted by the songs of males of their own species.

The species of periodical cicadas are *Magicicada septendecim, Magicicada cassini,* and *Magicicada septendecula.* They have no common names, but are sometimes referred to as simply *decim, cassini, and decula.* Each species has its own distinctive size, color pattern, song, and habitat preference, differences that tend to keep the three species apart and thus prevent hybridization between them. *Decim,* which is usually the most common species in northern broods, prefers mature upland forests and lays its eggs in the twigs of many different species of woody plants except for conifers other than juniper. In Connecticut, Chris Maier found this species laying its eggs in 49 species of trees and shrubs. *Cassini,* generally the most abundant species in southern broods, favors flood plain habitats and lays its eggs in such lowland trees as ashes, elms, and certain oaks. *Decula* is fairly abundant in a few broods but is usually the rarest of the three species and is altogether absent from some broods. Like *decim,* it occupies upland forests but is much fussier about where it lays its eggs, favoring mainly hickory and walnut trees.

Today most authorities believe that the three species that occur in 17-year broods are the same as the three that occur in 13-year broods. But some biologists consider the 13-year forms to be distinct species that are separate from the 17-year forms and give them different but corresponding names with a prefix that indicates the 13-year duration of their life cycle: *Magicicada tredecim, Magicicada tredecassini,* and *Magicicada tredecula.* The 13-year forms do not differ from their 17-year counterparts in any noticeable way, not in anatomy, song, choice of habitat, or other behaviors. Nevertheless, proponents of recognizing six rather than only three species of periodical cicadas argue that the 13- and 17-year forms can be legitimately designated as separate species because they are so widely separated in time and are thus reproductively isolated from each other.

But is this separation in time really enough to prevent significant mixing between 13-and 17-year forms? Perhaps not. After all, 13- and 17-year broods that occupy the same area do emerge together every 221 years. Although 221 years may seem like a very long time, it is only 13 generations of a 17-year brood and may not be long enough to keep the two forms so separated that they have evolved to become distinct species. And as Lewis Stannard wrote, contacts between 17- and 13-year broods are sometimes more frequent. In Illinois, for example, the area occupied by a 13-year brood (XIX) borders on the ranges of three different 17-year broods, one of which will coincide with the 13-year brood once every 65 years (5 generations) or once every 78 years (6 generations). Furthermore, Kathy Williams and Chris Simon reported that 17-year broods sometimes accelerate their development by exactly 4 years to become 13-year broods. In at least one instance a large part of a 17-year brood emerged 4 years early and joined a 13-year brood, and presumably interbred with them.

Charles Darwin agreed with the view that the 13-year and 17-year forms are not separate species in an 1868 letter to Benjamin Walsh, the first state entomologist of Illinois, and once Darwin's classmate at Trinity College, Cambridge University. This letter, in the collection of the Field Museum of Natural History in Chicago, was quoted in part by Dybas and Lloyd. In 1868 *decula* had not yet been discovered, but there was talk of designating the 13- and 17-year forms of *decim* and *cassini* as separate species. Darwin, reporting on a conversation with Asa Gray and Joseph Hooker, prominent naturalists of the day, wrote: "They thought that the 13- and 17-year forms ought not to be ranked as distinct species unless other differences besides the period of development could be discovered. They thought that the mere rarity of variability in such a point was not sufficient and I think I concur with them."

A male periodical cicada, as Alexander and Moore note, produces three different acoustic signals, each characteristic of its species, that differ in function and can be distinguished from each other because they sound quite different: the loud *disturbance squawk,* which I have already mentioned, a loud *calling song* that attracts both males and females, and a quieter *courtship song* that a male sings when he approaches a female or sometimes, presumably unknowingly, another male.

The three species of cicadas within either a 13- or 17-year brood are reproductively isolated from each other—thus maintaining their integrity and separateness—mainly because their calling songs are different and because they call at different times of the day. Corresponding 13- and 17-year forms do not differ from each other in these respects. The

songs of periodical cicadas are impossible to describe in words, but I can tell you that, as Alexander and Moore found, the calling songs of the three species are different enough in rhythm and pitch to be easily distinguished by the human ear. Furthermore, although there may be a little overlap between species, *decim* calls mainly in the morning, *cassini* mainly during the middle of the day, and *decula* mainly in the afternoon. The completeness of the separation between species is demonstrated by observations made by Alexander and Moore. They checked thousands of mating pairs in the field but did not find even one pair in which a male and a female of different species were coupled. However, mixed species couplings did occur when cicadas were kept in cages, presumably because the mechanisms that keep the species apart at least sometimes fail under conditions of close confinement.

❀ ❀ ❀ Since the calling song of periodical cicadas attracts males as well as females, many males of the same species will congregate in the same tree. Over a period of about 2 weeks, a congregation may grow to include hundreds or even thousands of males, forming a chorus in which they all call together, the individuals out of synchrony with each other so that they produce, as if with one voice, a loud and continuous ear-piercing drone. Many such choruses form within an area. Females are strongly attracted by a chorus of their own species, especially if it is large and outstandingly loud. But a female does not stay with the males for long. She mates soon after she arrives, and not long after separates from her mate and leaves to get on with the business of laying her eggs. Almost all matings take place within a chorus, and there is good reason to think that males that join a chorus are more successful at obtaining matings than are males that go it alone. Observations described to me by Tom Moore support this assumption by showing how totally committed males are to being a part of a chorus. First, if a male is removed from his chorus, he immediately returns to it. Second, males in a chorus are apparently stimulated to sing by the other singing males around them. Males in a chorus sing about 10 songs per minute, while lone males sing much less often, as seldom as 10 times per day.

Male cicadas in a chorus will court almost anything. Tom Moore tells me that they even tried to court small microphones he placed in a chorus tree. But females usually reject males that approach them, driving them away or even kicking them off the branch. Choosing a mate is largely the prerogative of the females, and most successful matings begin with a female approaching a male. The female flies to the chorus tree, and after she lands she approaches the male of her choice on foot. The male recognizes the female by her behavior—she does not sing and just stands still. Then he goes into a frenzy of courtship: sidling up to her, waggling his foreleg, often near her eye, and singing his courtship song. The courtship song of the male is characteristic of his species, differing markedly from the courtship songs of the other two species. It probably has two functions. It could help the female recognize the male as a member of her own species, and it could inform her of the quality of the male as a potential mate. Males that have an advanced fungus infection—poor mating prospects—sound different than do healthy males.

⚙ ⚙ ⚙ Inseminated females lay their eggs in rows of small cavities, egg nests, that they excavate with their ovipositor in the underside of a twig of a woody plant, usually a twig that is about the diameter of a lead pencil. The cicada's ovipositor is well suited for penetrating even fairly hard wood. It is long, rigid, strong, and has a sharp, serrated, sawlike tip. According to JoAnn White, a female does not lay eggs in a twig until she has judged it to be suitable by determining that it has the right size, taste, and texture. She gauges the diameter of a twig by grasping it with her middle legs. (Is it too thin or too thick?) She tastes the twig by rubbing or penetrating it with her mouthparts. (Does it contain chemical substances that might harm her eggs?) She judges the texture of the wood by probing it with her ovipositor. (Is it too hard to be easily penetrated?)

An egg nest is easily spotted because of the conspicuous tuft of light-colored splintered wood fibers that sticks out from its opening. White found that she could tell which species of periodical cicada had made an egg nest by its shape and structure and by the arrangement of the wood fibers associated with it. The average number of eggs placed in a nest is different for the three species. But any given nest may be empty,

contain only 1 or 2 eggs, or hold as many as 30. White writes that the number of eggs in a nest varies not only with the species of cicada but also with the species of plant in which the nest is located. Nests in the most favorable plants are likely to contain the maximum number of eggs, while those in unfavorable plants are likely to be abandoned before they have been filled, and may thus contain only a few eggs or none at all.

The eggs may hatch after as little as 6 weeks, but as Chris Maier reported, almost all the eggs laid by the 1979 contingent of brood II in Connecticut hatched during the month of August, from 9 to 10 weeks after they had been laid. As described by Ephraim Felt, state entomologist of New York from 1898 to 1928, "The recently hatched cicadas are slender grublike creatures, about 1/10 inch long. They are as lively as ants, and after running about on the tree for a short time, drop to the ground and bury themselves. Their strong forelegs are well adapted for digging, and are undoubtedly of great service in searching for the tender succulent rootlets on which they feed." After forming a feeding cell around a rootlet, the nymph inserts its piercing-sucking beak in the rootlet and begins to suck up the sap that will be its only food for the next 13 or 17 years.

Newly hatched as well as somewhat older but still small nymphs are relatively close to the soil surface, usually less than 9 inches beneath it. Their feeding cells are elongated and associated with very small roots. Larger nymphs are generally deeper in the soil and form roughly spheroidal feeding cells that are associated with somewhat larger roots. These larger nymphs are usually not more than 24 inches beneath the surface but have been found as far down as 42 inches.

Both adult and nymphal cicadas use their piercing-sucking beaks to feed on xylem sap from twigs or roots. Trees and most other plants have two different tissues that conduct sap from one part of the plant to another: the xylem, which carries sap upward from the roots to the stems and leaves through a system of narrow tubes, and the phloem, which carries sap downward from the leaves to the roots through another system of narrow tubes. Many insects with piercing-sucking beaks, among them aphids and scale insects, ingest phloem sap, a relatively nutritious food that is composed of 30 percent nutrients and only 70 percent water. But far fewer, among them cicadas and spittle bugs, ingest xylem sap, which is extremely dilute, consisting of about 99 percent water and less than 1 percent nutrients. Not only is xylem sap inferior to phloem sap as a food, but it is also much more difficult to obtain. Phloem sap is under positive pressure and is, therefore, easy to get out of the plant and does not even require the application of suction. If a feeding aphid is snipped away from its beak, leaving the severed beak embedded in the plant with its tip inserted in a phloem tube, sap continues to flow from the cut end of the beak. By contrast, xylem sap is under negative pressure because it is drawn upward as water transpires from the leaves. The cicadas can ingest xylem sap because the suction pumps at the base of their beaks are relatively huge and powerful in both the nymph and the adult. The aphids are restricted to phloem because their suction pumps are too small and weak to extract xylem.

❄ ❄ ❄   Compared to virtually all other insects, periodical cicadas grow extremely slowly. Just how slowly is vividly illustrated by comparing the rate at which a periodical cicada gains weight with the rate at which an insect that grows at a more usual pace gains weight. (This is most conveniently done if weights are expressed in milligrams. It takes 28,350 milligrams to make an ounce.) During its period of growth, the nymphal stage, a periodical cicada of the species *decim* gains about 1,000 mg, thus growing at a rate of about 0.16 mg per day if it is a 17-year individual or at about 0.21 mg per day if it is a 13-year individual. The larvae of the tobacco hornworm, the big green caterpillars that you may have found eating the leaves of your tomato plants, grow at a much faster rate. This caterpillar, the growing stage, gains about 7,000 mg in about 18 days. Its growth rate, adjusted for the sevenfold difference in weight between the cicada nymph and the caterpillar, is about 57 mg per day. Thus the caterpillar grows at a rate about 250 times greater than a 13-year *decim* and about 300 times greater than a 17-year *decim.*

The slow growth of periodical cicadas has been attributed to their total dependence on a nutrient-poor food. But that poor diet is probably no more than a minor factor in the retardation of their growth rate. After all, other insects that feed on xylem sap, including dog-day cicadas, grow much more rapidly and get even larger than do periodical cicadas. It seems probable, therefore, that—as maladaptive as it may seem at

first glance—periodical cicadas are genetically programmed to develop slowly. The disadvantages of slow growth are apparently outweighed by the escape from parasites and predators that periodical cicadas enjoy because they emerge as huge synchronized groups at intervals of many years.

Associating with Other Species

Lycid beetles crawling on blossoms of white clover

❁ ❁ ❁   Not only do members of the same species form groups, but in a few cases members of different species join together, often to their mutual benefit. Human society is really a mixed species association, although most of us don't look at it that way. Consider just the pets that we keep. Cats are welcome members of our households, and on many farms the barnyard cats earn their saucers of milk by destroying rats and mice. My neighborhood would seem incomplete without people walking the dog with pooper-scooper in hand. Although most of us think of dogs as nothing more than pets, they did, and sometimes still do, have practical roles in human affairs: as guards, as helpers in the hunt, and as herders of sheep and cattle. Dogs, the first animals domesticated by people, became associated with us over 12,000 years ago—probably as people came to rely for early warnings of approaching danger on the barks of wild dogs that hung around encampments scrounging for scraps.

❁ ❁ ❁   As surprising as it may seem, some insects keep domestic animals much as do humans. For example, cornfield ants, a common species in the corn-growing areas east of the Rocky Mountains, keep corn root aphids as permanent guests in their underground nests. The relationship is mutually beneficial, and the aphids could not survive without the ants. The nutritious honeydew produced by the sap-sucking aphids is an important part of the ants' diet, and the aphids are protected and provided with food by the ants. Honeydew, the liquid excrement of the aphids, is mostly phloem sap and contains a lot of sugar, but also some amino acids, the building blocks of proteins. Honeydew is eaten by many different kinds of insects, and in parts of the Mideast Arabs make candy from it. Some scholars have even suggested that honeydew was the manna from heaven that nourished the Israelites on their journey from Egypt to the promised land.

In the winter, the corn root aphids are all in the egg stage and are scrupulously cared for in the underground tunnels of the ants. As climatic conditions change, the ants move the eggs from place to place to keep them at favorable temperature and moisture levels. When the eggs hatch in the spring, the ants place the nearly helpless aphids on plant roots they find by tunneling through the soil. The aphids will suck sap from the roots of several grasses and broad-leaved weeds, but their favorite is corn, and the ants seem to know this. Robert L. Metcalf and Robert A. Metcalf reported an observation of cornfield ants moving their colony and their aphids along with it from a meadow of grass to a corn field about 50 yards away. The workers carried the larval and pupal ants and a large number of aphids, soon placing the latter on roots of the corn plants.

All aphids, the corn root aphid among them, are races of amazons that consort with males only during the last generation of the year. Throughout spring and summer there are only female corn root aphids, which reproduce parthenogenetically, without being fertilized by a male, and give live birth to their daughters. Most of the daughters are wingless and stay in the ants' nest to produce another generation. A few are winged and fly away. When they land they are likely to be discovered and adopted by another colony of cornfield ants. Otherwise they die. The last generation of parthenogenetic females produces a sexually reproducing generation of both wingless males and wingless females, which mate and lay the only eggs of the year.

⚙ ⚙ ⚙   Caterpillars of at least 10 families, as Matthew Bayliss and Naomi Pierce have pointed out, associate more or less closely with certain kinds of ants. Among them are some species of the family Riodinidae, the butterflies known as metalmarks and, especially familiar to those of us in the north temperate zone, some species of the family Lycaenidae, the gossamer-winged butterflies—mainly the dainty little ones known in the adult stage as blues.

In a very few cases the caterpillars are parasitic on the ants, living within their colonies and eating their larvae and pupae. But the relationships are usually mutualistic: both the caterpillars, which are usually plant feeders, and the ants benefit. The caterpillars produce nutritional secretions that are often an important part of the diet of the ants, and the ants, in return, do not attack the caterpillars and in many cases protect them from other predators and parasites.

Lycaenid caterpillars that associate with ants are variously adapted to placate and protect themselves from their ant associates. They have thick skins that protect their internal organs from the occasional misdirected bite of an ant. Many, if not all of them, as P. J. DeVries found, have small, sound-producing organs which they use to send apparently placating signals that are carried to the ants as vibrations of the substrate. And, as Konrad Fiedler and Ulrich Maschwitz noted, the caterpillars also secrete chemical substances that mimic the pheromones by which ants recognize and communicate with each other.

Some lycaenids simply placate ants, presumably to protect themselves, and do not reward them. But others, as I already mentioned, reward the ants by secreting nutrients that the ants consume. Many lycaenid caterpillars have an organ on the upper side of the abdomen that, at least in one species, is known to secrete mainly sugars in a concentration of as much as 19 percent by weight. This secretion contains only trace amounts of amino acids, the building blocks of proteins, but single-celled glands scattered over the surface of the skin produce larger amounts of amino acids.

Unlike the honeydew excreted by aphids, which is a cost-free by-product of feeding on the sugar-rich phloem sap of plants, the nutrient secretions of the leaf-eating caterpillars are produced at a steep price. These secretions are metabollically expensive because they must be actively manufactured or diverted from their essential role in growth and development.

At least some of the caterpillars are known to receive an important payback for their "gift" of costly nutrients, as is to be expected. Some lycaenid caterpillars have never been known to survive being discovered by predators unless they are attended by protecting ants. The protection is not complete, but Naomi Pierce and S. Easteal found that caterpillars of the silvery blue, a lycaenid widespread in North America, are from 4 to 12 times more likely to survive to the pupal stage when they are guarded by ants than when they are not.

Much more remains to be learned about how and how well ants protect caterpillars that provide them with food. But several fascinating cases are understood at least in broad outline. While some lycaenid caterpillars live exposed on their food plants and are tended by ants from nearby colonies, others are more closely associated with ants. Several plant-eating species live within an ant colony. Some live in ant nests made within plants and feed on the internal tissues of the plant. Others live within colonies of weaver ants, feeding on the leaves that the ants tie together to form their nests. Certain Australian lycaenid caterpillars climb up into a tree to feed on leaves during the night, but descend to hide throughout the daylight hours in earthen corrals that their attendant ants build at the base of the tree.

✿ ✿ ✿ A few species of ants keep other ants as domestic animals. Entomologists generally call this practice "slavery," but in its usual sense, this word denotes the enslavement of members of one's own species, a rare practice among animals, known to occur only in humans and two species of ants. With these two exceptions, the "slaves" held by ants are really domestic animals, just as are people's dogs, goats, cattle, and horses.

In their everyday lives, the amazon ants of Europe and North America are completely dependent upon the workers of other species of ants that are their household "slaves." They take over the nest of a "slave" species, but do not and apparently are not capable of enlarging it, cleaning it, feeding themselves, or caring for their own offspring. As William Morton Wheeler wrote, when the amazons are not stolidly idle, they groom their bodies or beg their "slaves" for food. They do, however, become determinedly active and aggressive when it becomes necessary to make a raid to replenish the population of "slaves." As Wheeler wrote: "The ants leave the nest very suddenly and assemble about the entrance if they are not, as sometimes happens, pulled back and restrained by their slaves. Then they move out in a compact column with feverish haste . . . On reaching the nest to be pillaged, they do not hesitate but pour into it at once in a body, seize the brood, rush out again and make for home." If they are attacked by the defending workers, they kill them by piercing their heads or thoraxes with their wickedly sharp, sickle-shaped mandibles. After the raiders return to their home nest, the "slaves" raise the newly captured brood, larval and pupal workers, to maturity. By the time they become adults, the new "slave" workers have accepted the colony of their masters as their own.

✦ ✦ ✦   Both in temperate and in tropical areas, flocks consisting of several different species of birds, sometimes a dozen or more, can be seen flitting through trees and shrubs as they hunt for insects. How or if the birds benefit from such interspecific associations was largely a subject of speculation until Kimberly Sullivan studied the mixed species feeding flocks that commonly occur in the eastern United States and southern Canada. These flocks may include insectivorous birds of several species: downy woodpeckers, chickadees, tufted titmice, sapsuckers, nuthatches, brown creepers, kinglets, and yellow-rumped warblers. She demonstrated that downy woodpeckers benefit from joining a mixed species flock because they can devote less time to watching for predators and spend more time searching for food and eating than when they are alone. The other species in the flock presumably benefit in the same way.

Sullivan did an experiment in the field which showed that these woodpeckers recognize the alarm calls of other members of the flock as warnings of an approaching predator. They also assure and reassure themselves that the other members of the flock are actually present and vigilant by listening for the social calls—as opposed to alarm calls—that announce their presence to other members of their own species. Sullivan had noticed that foraging downy woodpeckers often cock their heads as if to watch for predators. In her experiment, she played to lone downy woodpeckers foraging in the wild the recorded social calls of birds that they often flock with, black-capped chickadees and tufted titmice. As a result, these lone woodpeckers cocked their heads less often and spent more time feeding, although other birds were not actually present. But in control experiments, lone downy woodpeckers did not cock their heads less often in response to the recorded social calls of seed-eating birds that they seldom if ever flock with, American tree sparrows, dark-eyed juncos, and American goldfinches.

❄ ❄ ❄   Like the birds just discussed, certain insects enhance their protection against predators by coming together in mixed species feeding groups. Among them are lycid beetles. Gastronomically speaking, lycids are an obnoxious lot. If they are disturbed by a person, or presumably by a bird or some other predator, they eject a foul-smelling fluid from their hind legs. Their noxiousness is attested to by the fact that many animals refuse to eat them: lizards, birds, mammals, and even some insects, including ants, mantises, and wasps. Their odor and conspicuousness warn away would be predators that have had a previous experience with a lycid. These beetles are, as you read in an earlier chapter, Müllerian mimics of each other. Different species of lycids, especially ones that occur in the same area, have evolved such a close resemblance to each other that predators probably can't tell them apart. Each individual of this Müllerian complex gains, since it is less likely to become an object lesson for a predator because the population pool from which naive predators take their first—and probably last—lycid is larger than it would otherwise be.

At the Southwestern Research Station in the Chiricahua Mountains of southeastern Arizona, E. G. Linsley and two co-investigators found that four species of lycids, all Müllerian mimics of each other, aggregate on the same individual plants to feed on pollen and nectar. These mixed species groups, with all individuals in close proximity, probably intensify the visual and odorous warnings of these noxious beetles, and thereby enhance their protection against predators.

Caterpillars have been known to form mixed species groups. In 1900, the British entomologist J. W. Tutt cited a report that two unrelated species of caterpillars sometimes occupy the same silken tent, European tent caterpillars and the caterpillars of the European gold-tail moth. The two are so distantly related that they are placed in different families. To quote Tutt, a German observer "once saw larvae [tent caterpillars] busily engaged in repairing and enlarging a 'gold tail' web, which they occupied together with the 'gold-tails,' feeding with them and accepting them as members of the same family."

What benefits could these two unrelated species get from living to-gether? There are, of course, all the advantages associated with living in a large group rather than a smaller one. But there is another possible ad-vantage. The body of a gold-tail caterpillar bears urticating hairs that sting like nettles, but tent caterpillars have no such protection and are likely to be protected from birds and other predators by associating with the better protected gold-tail caterpillars.

⚙ ⚙ ⚙ As you have read, being a part of a group—be it a cohesive unit or only a loose assemblage—can benefit an individual in several ways: A group may be better able to cope with adverse weather than can a lone individual. Group defenses against predators or parasites may be more effective than individual defenses. Groups may be more efficient at finding food and subduing it if it fights back, be it an animal or a plant. Finally, associating with many others of the same species usually simplifies the problem of finding a mate.

But group living can have serious disadvantages. Some predators and parasites target groups, as do the ladybird beetles and a few other pred-ators that live within colonies of aphids. The diseases of insects—caused by viruses, bacteria, fungi, and other pathogens—may spread more rap-idly within a group than between widely spaced individuals. Gypsy moth caterpillars, for example, can be killed by a viral disease. When they are not numerous and relatively scattered, only a few succumb to the disease, but when their population burgeons and they are crowded, the disease becomes rampant and the caterpillar population plummets. Insects that utilize small and scattered food sources may be better off going it alone rather than competing for food with others in a group. Contrast, for example, two species of hornworms, the caterpillar stage of hawk moths. Tomato hornworms do not form groups. They feed on small herbaceous plants that are not large enough to support more than a few of the ravenous caterpillars and that under natural conditions may be widely scattered. But their relatives, the catalpa hornworms, can afford to occur in huge groups because they feed on catalpa trees, which bear enough leaves to feed hundreds of these large caterpillars.

Because of these and other disadvantages, many insects—probably most of them—cannot afford to live in groups. But for some species, the

disadvantages are more than offset by the advantages, making it possible for them to reap the benefits of close association with others. For example, although an individual in a group may be more likely to contract a disease than is a lone individual, this greater risk may be compensated for by an increase in foraging efficiency or in the probability of finding a mate.

Selected Readings

Acknowledgments

Index

Selected Readings

Strength in Numbers

Bourke, A. F. G. 1995. *Social Evolution in Ants.* Princeton: Princeton University Press.

Byers, J. A. 1995. Host-tree chemistry affecting colonization in bark beetles. In R. T. Cardé and W. J. Bell (eds.), *Chemical Ecology of Insects 2.* New York: Chapman and Hall.

Calahane, V. H. 1947. *Mammals of North America.* New York, MacMillan.

Dybas, H. S., and D. D. Davis. 1962. A population census of seventeen-year periodical cicadas (Homoptera: Cicadidae: *Magicicada). Ecology* 43: 432–444.

Frisch, Karl von. 1968. *The Dance Language and Orientation of Bees.* Trans. L. E. Chadwick. Cambridge, Mass.: Harvard University Press.

———1971. *Bees: Their Vision, Chemical Senses, and Language.* Ithaca: Cornell University Press.

Hailman, J. P., K. J. McGowan, and G. E. Woolfenden. 1994. Role of helpers in the sentinel behaviour of the Florida scrub jay (*Aphelocoma c. coerulescens). Ethology* 97: 119–140.

Hamilton, W. D. 1963. The evolution of altruistic behavior. *American Naturalist* 97: 354–356.

———1964. The genetical evolution of social behavior II. *Journal of Theoretical Biology* 7: 17–52.

———1971. Geometry of the selfish herd. *Journal of Theoretical Biology* 31: 295–311.

Haviland, M. D. 1926. *Forest, Steppe, and Tundra.* Cambridge: Cambridge University Press.

Henry, C. S. 1972. Eggs and rapagula of *Ululodes* and *Ascaloptynx* (Neuroptera: Ascalaphidae): a comparative study. *Psyche* 79: 1–22.

Hölldobler, B., and E. O. Wilson. 1990. *The Ants.* Cambridge, Mass.: Harvard University Press.

———1994. *Journey to the Ants.* Cambridge, Mass.: Harvard University Press.

Hudleston, J. A. 1958. Some notes on the effects of bird predators on hopper bands of the desert locust (*Schistocerca gregaria* Forskal). *Entomologist's Monthly Magazine* 94: 210–214.

Jolivet, P., and T. J. Hawkeswood. 1995. *Host-plants of Chrysomelidae of the World.* Leiden: Backhuys.

López, E. R., L. C. Roth, D. N. Ferro, D. Hosmer, and A Mafra-Neto. 1997. Behavioral ecology of *Myiopharus doryphorae* (Riley) and *M. abberans* (Townsend), tachinid parasitoids of the Colorado potato beetle. *Journal of Insect Behavior* 10: 49–78.

Mayr, E. 1970. *Populations, Species, and Evolution.* Cambridge, Mass.: Harvard University Press.

McGowan, K. J., and G. E. Woolfenden. 1989. A sentinel system in the Florida scrub jay. *Animal Behaviour* 37: 1000–1006.

Michener, C. D. 1974. *The Social Behavior of the Bees.* Cambridge, Mass.: Harvard University Press.

Ross, K. G., and R. W. Matthews. 1991. *The Social Biology of Wasps.* Ithaca: Comstock Publishing Associates.

Seeley, T. D. 1995. *The Wisdom of the Hive.* Cambridge, Mass.: Harvard University Press.

Sober, E., and D. S. Wilson. 1998. *Unto Others: The Evolution and Psychology of Unselfish Behavior.* Cambridge, Mass.: Harvard University Press.

Tener, J. S. 1954. *A Preliminary Study of the Musk-oxen of Fosheim Peninsula, Ellesmere Island, N.W.T.* Wildlife Management Bulletin of the Canadian Wildlife Service, Ottawa, series 1, no. 9.

Wilson, E. O. 1971. *The Insect Societies.* Cambridge, Mass.: Harvard University Press.

———1975. *Sociobiology: The New Synthesis.* Cambridge, Mass.: Harvard University Press.

Wynne-Edwards, V. C. 1962. *Animal Dispersion in Relation to Social Behavior.* Edinburgh: Oliver and Boyd.

Bunches of Beetles

Carnes, E. K. 1912. Collecting ladybirds (Coccinellidae) by the ton. *Monthly Bulletin of the California State Commission of Horticulture* 1: 71–81.

Davidson, W. M. 1919. The convergent ladybird beetle (*Hippodamia convergens* Guerin) and the barley-corn aphis (*Aphis maidis* Fitch). *The Monthly Bulletin of the California State Commission of Horticulture* 8: 23–26.

———1924. Observations and experiments on the dispersion of the convergent lady-beetle (*Hippodamia convergens* Guérin) in California. *Transactions of the American Entomological Society* 50: 163–175.

DeBach, P., and K. S. Hagen. 1964. Manipulation of entomophagous species. In P. DeBach (ed.), *Biological Control of Insect Pests and Weeds.* New York, Reinhold Publishing Corporation.

Doutt, R. L. 1964. The historical development of biological control. In P. Debach (ed.), *Biological Control of Insect Pests and Weeds.* New York: Reinhold Publishing Corporation.

Garman, P. 1936. Control of apple aphids with California lady beetles. *Connecticut Agricultural Experiment Station Bulletin* 383: 356–357.

Obrycki, J. J., and T. J. Kring. 1998. Predaceous coccinellidae in biological control. *Annual Review of Entomology* 43: 295–321.

Packard, C. M., and Roy E. Campbell. 1926. The pea aphid as an alfalfa pest in California. *Journal of Economic Entomology* 19: 752–761.

Warding Off Predators

Allen, J. L., K. Schulze-Kellman, and G. J. Gamboa. 1982. Clumping patterns during overwintering in the paper wasp, *Polistes exclamans:* effects of relatedness. *Journal of the Kansas Entomological Society,* 55: 97–100.

Barlow, N. (ed.). 1958. *The Autobiography of Charles Darwin, 1809–1882.* London: Collins

Bates, H. W. 1862. Contributions to an insect fauna of the Amazon Valley, Lepidoptera: Heliconidae. *Transactions of the Linnaean Society, Zoology* 23: 95–566.

Beebe, W. 1918. The convict trail. *Atlantic Monthly,* September.

Benson, W. 1972. Natural selection for Müllerian mimicry in *Heliconius erato* in Costa Rica. *Science* 176: 936–939.

Boyden, T. C. 1976. Butterfly palatability and mimicry: experiments with *Ameiva* lizards. *Evolution* 30: 73–81.

Brindley, H. H. 1906. The procession of *Cnethocampa pinivora*, Treitschke. *Proceedings of the Cambridge Philosophical Society* 14: 98–104.

Brower, L. P., J. V. Z. Brower, and C. T. Collins. 1963. Experimental studies of mimicry. 7. Relative palatability and Müllerian mimicry among neotropical butterflies of the subfamily Heliconiinae. *Zoologica* 48: 65–84.

Edwards, J. S. 1961. Spitting as a defensive mechanism in a predatory reduviid. *Proceedings of the Eleventh International Congress of Entomology* 3: 259–263.

Eisner, T., and F. C. Kafatos. 1962. Defense mechanisms of arthropods. X. A pheromone promoting aggregation in an aposematic distasteful insect. *Psyche* 69: 53–61.

Frazer, J. F. D., and M. Rothschild. 1960. Defence mechanisms in warningly-coloured moths and other insects. *Proceedings of the Eleventh International Congress of Entomology* 3: 249–256.

Hudson, W. H. 1912. *The Naturalist in La Plata*, 5th ed. London: J. M. Dent and Sons.

Janzen, D. H. (ed.). 1983. *Costa Rican Natural History.* Chicago: University of Chicago Press.

Jones, F. M. 1930. The sleeping heliconias of Florida. *Natural History* 30: 638–644.

Linsenmaier, W. 1972. *Insects of the World.* Trans. L. E. Chadwick. New York: McGraw-Hill.

Mallis, A. 1971. *American Entomologists.* New Brunswick, N.J.: Rutgers University Press.

Matthews, R. W., and J. R. Matthews. 1978. *Insect Behavior.* New York: John Wiley and Sons.

Morrow, P. A., T. E. Bellas, and T. Eisner. 1976. *Eucalyptus* oils in the defensive oral discharge of Australian sawfly larvae (Hymenoptera: Pergidae). *Oecologia* 24: 193–206.

Mortensen, T. 1918. Papers from Dr. Thomas Mortensen's Pacific expedition. I. Observations on protective adaptations and habits, mainly in marine animals. *Videnskabelige Meddelelser fra Dansk naturhistorisk Forening i Kjøbenhavn* 69: 57–96.

Moss, A. M. 1920. Sphingidae of Para, Brazil. *Novitates Zoologicae* 27: 333–424.

Müller, F. 1879. *Ituna* and *Thyridis;* a remarkable case of mimicry in butterflies. Trans. R. Meldola. *Proceedings of the Entomological Society of London* 27: xx-xxix.

Rau, P. 1941. The swarming of *Polistes* wasps in temperate regions. *Annals of the Entomological Society of America* 34: 580–584.

Rau, P., and N. Rau. 1916. The sleep of insects: an ecological study. *Annals of the Entomological Society of America* 9: 227–274.

Rothschild, M. 1983. *Dear Lord Rothschild: Birds, Butterflies, and History.* Glenside, Penn.: Balaban Publishers.

Schmidt, J. O. 1990. Hymenopteran venoms: striving toward the ultimate defense against vertebrates. In D. L. Evans and J. O. Schmidt (eds.), *Insect Defenses.* Albany: State University of New York Press.

Sillén-Tullberg, B. 1985. Higher survival of an aposematic than of a cryptic form of a distasteful bug. *Oecologia* 67: 411–415.

Sillén-Tullberg, B., and O. Leimar. 1988. The evolution of gregariousness in distasteful insects as a defense against predators. *American Naturalist* 132: 723–734.

Spencer, K. C. 1988. Chemical mediation of coevolution in the *Passiflora-Heliconius* interaction. In K. C. Spencer (ed.), *Chemical Mediation of Coevolution.* San Diego: Academic Press.

Starr, C. K. 1990. Holding the fort: colony defense in some primitively social wasps. In D. L. Evans and J. O. Schmidt (eds.), *Insect Defenses.* Albany: State University of New York Press.

Turner, J. R. G. 1975. Communal roosting in relation to warning colour in two heliconiine butterflies (*Nymphalidae*). *Journal of the Lepidopterorists' Society* 29: 221–226.

Millions of Monarchs

Barker, J. F., and W. S. Herman. 1976. Effect of photoperiod and temperature on reproduction of the monarch butterfly, *Danaus plexippus. Journal of Insect Physiology* 22: 1565–1568.

Brower, J. V. Z. 1958. Experimental studies of mimicry in some North American butterflies. Part I. The monarch, *Danaus plexippus,* and viceroy, *Limenitis archippus archippus. Evolution* 12: 32–47.

———1960. Experimental studies of mimicry. IV. The reactions of starlings to different proportions of models and mimics. *The American Naturalist* 94: 271–282.

Brower, L. P. 1969. Ecological chemistry. *Scientific American* 220: 22–30.

———1992. The current status of butterfly royalty. *Terra* 30: 4–15.

———1995. Understanding and misunderstanding the migration of the monarch butterfly (Nymphalidae) in North America: 1857–1995. *Journal of the Lepidopterists' Society* 49: 304–385.

Brower, L. P., and W. H. Calvert. 1985. Foraging dynamics of bird predators on overwintering monarch butterflies in Mexico. *Evolution* 39: 852–868.

Brower, L. P., W. H. Calvert, L. E. Hedrick, and J. Christian. 1977. Biological observations on an overwintering colony of monarch butterflies (*Danaus plexippus, Danaidae*) in Mexico. *Journal of the Lepidopterists' Society* 31: 232–242.

Calvert, W. H., L. E. Hedrick, and L. P. Brower. 1979. Mortality of the monarch butterfly (*Danaus plexippus* L.): avian predation at five overwintering sites in Mexico. *Science* 204: 847–851.

Curio, E. 1976. *The Ethology of Predation.* Berlin: Springer.

Fink, L. S., and L. P. Brower. 1981. Birds can overcome the cardenolide defence of monarch butterflies in Mexico. *Nature* 291: 67–70.

Gibo, D. L., and J. A. McCurdy. 1993. Lipid accumulation by migrating monarch butterflies (*Danaus plexippus* L.). *Canadian Journal of Zoology* 71: 76–82.

Hill, H. F., Jr., A. M. Wenner, and P. H. Wells. 1976. Reproductive behavior in an overwintering aggregation of monarch butterflies. *American Midland Naturalist* 95: 10–19.

Losey, J. E., L. S. Rayor, and Maureen Carter. 1999. Transgenic pollen harms monarch larvae. *Nature* 399: 214.

Malcolm, S. B. 1995. Milkweeds, monarch butterflies, and the ecological significance of cardenolides. *Chemoecology* 5/6: 101–117.

Masters, A. R., S. B. Malcolm, and L. P. Brower. 1988. Monarch butterly (*Danaus plexippus*) thermoregulatory behavior and adaptations for overwintering in Mexico. *Ecology* 69: 458–467.

Perez, S. M., O. R. Taylor, and R. Jander. 1997. A sun compass in monarch butterflies. *Nature* 387: 29.

———1999. Monarch butterflies (*Danaus plexippus*) are disoriented by a strong magnetic pulse. *Naturwissenschaften* 86: 140–143.

Pliske, T. E. 1975. Courtship behavior of the monarch butterfly, *Danaus plexippus* L. *Annals of the Entomological Society of America* 68: 143–151.

Ritland, D. B. 1994. Variation in palatability of Queen butterflies (*Danaus gilippus*) and implications regarding mimicry. *Ecology* 75: 732–746.

Ritland, D. B., and L. P. Brower. 1991. The viceroy butterfly is not a Batesian mimic. *Nature* 350: 497–498.

Urquhart, F. A. 1957. *A Discussion of Batesian Mimicry as Applied to the Monarch and Viceroy Butterflies.* Toronto: University of Toronto Press.

———1960. *The Monarch Butterfly.* Toronto: University of Toronto Press.

———1976. Found at last: the monarch's winter home. *National Geographic* 150 (August): 161–173.

Urquhart, F. A., and N. R. Urquhart. 1976. The overwintering site of the eastern population of the monarch butterfly (*Danaus p. plexippus;* Danaidae) in southern Mexico. *Journal of the Lepidopterists' Society* 30: 153–158.

Wassenaar, L. I., and K. A. Hobson. 1998. Natal origins of migratory monarch butterflies at wintering colonies in Mexico: new isotopic evidence. *Proceedings of the National Academy of Sciences* 95: 15436–15439.

Teams of Tent Caterpillars

Bent, A. C. 1940. *Life Histories of North American Cuckoos, Goatsuckers, Hummingbirds, and their Allies.* Washington, D.C.: Smithsonian Institution Bulletin 176.

Fitzgerald, T. D. 1993. Sociality in caterpillars. In N. E. Stamp and T. M. Casey (eds.), *Caterpillars.* New York: Chapman and Hall.

————1995. *The Tent Caterpillars.* Ithaca: Cornell University Press.

Fitzgerald, T. D., T. Casey, and B. Joos. 1988. Daily foraging schedule of field colonies of the eastern tent caterpillar, *Malacosoma americanum. Oecologia* 76: 574–578.

Fitzgerald, T. D., and D. E. Willer. 1983. Tent building behavior of the eastern tent caterpillar *Malacosoma americanum* (Lepidoptera: Lasiocampidae). *Journal of the Kansas Entomological Society* 56: 20–31.

Forbush, E. H. 1927. *Birds of Massachusetts,* vol. 2. Boston: Massachusetts Department of Agriculture.

Futuyma, D. J., and S. S. Wasserman. 1981. Food plant specialization and feeding efficiency in the tent caterpillars *Malacosoma disstria* and *M. americanum. Entomologia Experimentalis et Applicata* 30: 106–110.

Myers, J. H., and J. N. M. Smith. 1978. Head flicking by tent caterpillars, a defensive response to parasite sounds. *Canadian Journal of Zoology* 56: 1628–1631.

Peterson, S. C. 1986. Breakdown products of cyanogenesis, repellancy and toxicity to predatory ants. *Naturwissenschaften* 73: 627–628.

Peterson, S. C., N. D. Johnson, and J. L. LeGuyader. 1987. Defensive regurgitation of allelochemicals derived from host cyanogenesis by eastern tent caterpillars. *Ecology* 68: 1268–1272.

Rissing, S. W., and G. B. Pollock. 1986. Social interaction among pleometrotic queens of *Veromessor pergandei* (Hymenoptera: Formicidae) during colony foundation. *Animal Behaviour* 34: 226–233.

Rissing, S. W., G. B. Pollock, M. R. Higgins, R. H. Hagen, and D. R. Smith. 1989. Foraging specialization without relatedness or dominance among co-founding ant queens. *Nature* 338: 420–422.

Willson, M. F. 1983. *Plant Reproductive Ecology.* New York: John Wiley and Sons.

Controlling the Climate

Denlinger, D. L. 1994. The beetle tree. *American Entomologist* 40: 168–171.

Eisner, T., A. F. Kluge, J. E. Carrel, and J. Meinwald. 1971. Defense of a phalangid: liquid repellent administered by leg dabbing. *Science* 173: 650–652.

Fitzgerald, T. D. 1995. *The Tent Caterpillars.* Ithaca: Cornell University Press.

Lindauer, M. 1967. *Communication among Social Bees.* New York: Atheneum.

Lüscher, M. 1961. Air-conditioned termite nests. *Scientific American* 205 (July): 138–145.

Michener, C. D. 1974. *The Social Behavior of the Bees.* Cambridge, Mass.: Harvard University Press.

Tanaka, S., H. Wolda and D. L. Denlinger. 1988. Group size affects the metabolic rate of a tropical beetle. *Physiological Entomology* 13: 239–241.

Wagner, H. O. 1954. Massenansammlungen von Weberknechten in Mexiko. (Mass aggregations of daddy longlegs in Mexico). *Zeitschrift für Tierpsychologie* 11: 349–352.

Wellington, W. G. 1974. Tents and tactics of caterpillars. *Natural History* 79 (January): 64–72.

Wigglesworth, V. B. 1972. *The Principles of Insect Physiology.* London: Chapman and Hall.

Wilson, E. O. 1971. *The Insect Societies.* Cambridge, Mass.: Harvard University Press.

Yoder, J. A., D. L. Denlinger, and H. Wolda. 1992. Aggregation promotes water conservation during diapause in the tropical fungus beetle, *Stenotarsus rotundus*. *Entomologia Experimentale et Applicata* 63: 203–205.

Coinciding with Resources

Borg, A. F., and W. R. Horsfall. 1953. Eggs of floodwater mosquitoes. II. Hatching stimulus. *Annals of the Entomological Society of America* 46: 472–478.

Branson, T. F., and J. L. Krysan. 1981. Feeding and oviposition behavior and life cycle strategies of *Diabrotica:* an evolutionary view with implications for pest management. *Environmental Entomology* 10: 826–831.

Chiang, H. C. 1965. Survival of northern corn rootworm eggs through one and two winters. *Journal of Economic Entomology* 58: 470–472.

Frisch, K. von. 1953. *The Dancing Bees.* Trans. Dora Ilse. New York: Harcourt, Brace, and World.

Harrington, B. 1996. *The Flight of the Red Knot.* New York: W. W. Norton.

Horsfall, W. R., H. W. Fowler, Jr., L. J. Moretti, and J. R. Larsen. 1973. *Bionomics and Embryology of the Inland Floodwater Mosquito,* Aedes vexans. Urbana: University of Illinois Press.

Krysan, J. L., J. J. Jackson, and A. C. Lew. 1984. Field termination of egg diapause in *Diabrotica* with new evidence of extended diapause in *D. barberi* (Coleoptera: Chrysomelidae). *Environmental Entomology* 13: 1237–1240.

Krysan, J. L., and T. A. Miller (eds.). 1986. *Methods for the Study of Pest Diabrotica.* New York: Springer-Verlag.

Levine, E., H. Oloumi-Sadeghi, and J. R. Fisher. 1992. Discovery of multiyear diapause in Illinois and South Dakota northern corn rootworm (Coleoptera: Chrysomelidae) eggs and incidence of the prolonged diapause trait in Illinois. *Journal of Economic Entomology* 95: 262–267.

Mangelsdorf, P. C. 1974. *Corn.* Cambridge, Mass.: Harvard University Press.

Seeley, T. D. 1985. *Honey Bee Ecology.* Princeton: Princeton University Press.

———1995. *The Wisdom of the Hive: The Social Physiology of Honey Bee Colonies.* Cambridge, Mass.: Harvard University Press.

Shuster, C. M., Jr. 1982. A pictorial review of the natural history and ecology of the horseshoe crab *Limulus polyphemus*, with reference to other Limulidae. *In* J. Bonaventura, C. Bonaventura, and S. Tesh (eds.), *Physiology and Biology of Horseshoe Crabs.* New York: Alan R. Liss.

Ward, D. P. 1992. *On Methuselah's Trail.* New York: W. H. Freeman.

Webster, F. M. 1895. On the probable origin, development, and diffusion of North American species of the genus *Diabrotica. Journal of the New York Entomological Society* 3: 158–166.

Subduing Food

Balduf, W. V. 1939. *The Bionomics of Entomophagous Insects,* Part II. St. Louis: John S. Swift.

Bongers, J., and W. Eggerman. 1971. Der Einfluss des Subsozialverhaltens der spezialisierten Samensauger *Oncopeltus fasciatus* Dall. und *Dysdercus fasciatus* Sign. auf ihre Ernährung. (Influence on their nutrition of the subsocial behavior of the specialized seed suckers *Oncopeltus fasciatus* Dall. and *Dysdercus fasciatus* Sign.) *Oecologia* 6: 293–302.

Burgess, J. W. 1976. Social spiders. *Scientific American* 234: 101–106.

Buskirk, R. E. 1981. Sociality in the Arachnida. In H. R. Herman (ed.), *Social Insects,* vol. II. New York: Academic Press.

Gabritschevsky, E. 1927. Experiments on color changes and regeneration in the crab spider, *Misumena vatia. Journal of Experimental Zoology* 47: 251–267.

Ghent, A. W. 1955. Oviposition behaviour of the jack-pine sawfly, *Neodiprion americanus banksianae* Roh., as indicated by an analysis of egg clusters. *The Canadian Entomologist* 87: 229–238.

———1960. A study of the group-feeding behaviour of larvae of the jack pine sawfly, *Neodiprion pratti banksianae* Roh. *Behaviour* 16: 110–148.

Gilbert, L. E. 1971. Butterfly-plant coevolution: has *Passiflora adenopoda* won the selectional race with Heliconiine butterflies? *Science* 172: 585–586.

Hölldobler, B., and E. O. Wilson. 1990. *The Ants.* Cambridge, Mass.: Harvard University Press.

Keen, F. P. 1952. Bark beetles in forests. In A. Stefferud (ed.), *Insects: The U.S.D.A. Yearbook of Agriculture, 1952.*

Ralph, C. P. 1976. Natural food requirements of the large milkweed bug, *Oncopeltus fasciatus* (Hemiptera: Lygaeidae), and their relation to gregariousness and host plant morphology. *Oecologia* 26: 157–175.

Rathke, B. J, and R. W. Poole. 1975. Coevolutionary race continues: butterfly adaptation to plant trichomes. *Science* 187: 175–176.

Schneirla, T. C. 1956. The army ants. In *Annual Report of the Board of Regents of the Smithsonian Institution for 1955,* pp. 379–406.

Schowalter, T. D., and G. M. Filip (eds.). 1993. *Beetle-Pathogen Interactions in Conifer Forests.* London: Academic Press.

Weis, A. E., and M. R. Berenbaum. 1989. Herbivorous insects and green plants. In W. G. Abrahamson (ed.), *Plant-Animal Interactions.* New York: McGraw-Hill.

Wilson, E. O. 1971. *The Insect Societies.* Cambridge, Mass.: Harvard University Press.

Legions of Locusts

Ashall, C., and P. E. Ellis. 1962. *Studies on Numbers and Mortality in Field Populations of the Desert Locust.* Anti-Locust Bulletin 38. London: Anti-Locust Research Centre.

Bhatia, D. R., and H. L. Sika. 1956. Some striking cases of food preference by the desert locust (*Schistocerca gregaria* Forsk.) *The Indian Journal of Entomology* 18: 205–210.

Carruthers, G. T. 1889. Locusts in the Red Sea. *Nature* 41: 53.

Chapman, R. F. 1976. *A Biology of Locusts.* London: Edward Arnold.

Ellis, P. E. 1959. Learning and social aggregation in locust hoppers. *Animal Behaviour* 7: 91–106.

Gillett, S. D. 1988. Solitarization in the desert locust, *Schistocerca gregaria* (Forskål) (Orthoptera: Acrididae). *Bulletin of Entomological Research* 78: 623–631.

Lockwood, J. A., R. A. Nunamaker, R. E. Pfadt, and L. D. DeBrey. 1990. Grasshopper Glacier: characterization of a vanishing biological resource. *American Entomologist* 36: 18–27.

Pener, M. P. 1991. Locust phase polymorphism and its endocrine relations. *Advances in Insect Physiology* 23: 1–79.

Riley, C. V. 1877. *The Locust Plague in the United States.* Chicago: Rand McNally.

———1878. Several chapters in *The First Annual Report of the United States Entomological Commission.* Washington, D.C.: U.S. Government Printing Office.

Sabrosky, C. W. 1952. How many insects are there? In A. Stefferud (ed.), *Insects: The U.S.D.A. Yearbook of Agriculture, 1952.*

Southey, R. 1909. *Poems of Robert Southey.* Ed. Maurice H. Fitzgerald. London: Oxford University Press.

Uvarov, B. 1921. A revision of the genus *Locusta*, L. (=*Pachystylus*, Fieb.), with a new theory as to the periodicity and migrations of locusts. *Bulletin of Entomological Research* 12: 135–163.

———1966. *Grasshoppers and Locusts*, vol. 1. London: Cambridge University Press.

———1977. *Grasshoppers and Locusts*, vol. 2. London: Centre for Overseas Pest Research.

Williams, C. B. 1958. *Insect Migration.* London: Collins.

People and Insect Plagues

Bell, D. N. 1992. *Wholly Animals: A Book of Beastly Tales.* Kalamazoo, Mich.: Cistercian Press.

Buck, P. S. 1931. *The Good Earth.* New York: John Day.

Christian, W. A., Jr. 1981. *Local Religion in Sixteenth-Century Spain.* Princeton: Princeton University Press.

Evans, E. P. 1906. *The Criminal Prosecution and Capital Punishment of Animals.* London: William Heinemann.

Evans, H. E. 1966. *Life on a Little-known Planet.* New York: E. P. Dutton.

Harpaz, I. 1973. Early entomology in the Middle East. In R. F. Smith, T. E. Mittler, and C. N. Smith (eds.), *History of Entomology.* Palo Alto, Calif.: Annual Reviews.

Packard, C. M., P. Luginbill, Sr., and C. Benton. 1937. *How to Fight the Chinch Bug.* U.S.D.A. Farmer's Bulletin 1780.

Pliny the Elder. 1856. *The Natural History of Pliny.* Ed. and trans. J. Bostock and H. T. Riley. London: Henry G. Bohn.

Swain, R. B. 1980. *Field Days.* Harmondsworth, Eng.: Penguin Books.

Winston, Mark L. 1997. *Nature Wars: People vs. Pests.* Cambridge, Mass.: Harvard University Press.

Finding a Mate

Akre, R. D., A. Greene, J. F. MacDonald, P. J. Landolt, and H. G. Davis. 1980. *Yellowjackets of America North of Mexico.* U.S.D.A. Agricultural Handbook 552.

Berenbaum, M. R. 1993. *Ninety-nine More Maggots, Mites, and Munchers.* Urbana: University of Illinois Press.

Buck, J. B. 1937. Flashing of fireflies in Jamaica. *Nature* 139: 801.

———1938. Synchronous rhythmic flashing of fireflies. *Quarterly Review of Biology.* 13: 301–314.

Buck, J. B., and E. Buck 1976. Synchronous fireflies. *Scientific American* 234: 74–85.

Chapman, J. A. 1954. Studies on summit-frequenting insects in western Montana. *Ecology* 35: 41–49.

Cott, H. B. 1957. *Adaptive Coloration in Animals.* London: Methuen.

Dethier, V. G. 1992. *Crickets and Katydids, Concerts and Solos.* Cambridge, Mass.: Harvard University Press.

Dodge, H. R., and J. M. Seago. 1954. Sarcophagidae and other Diptera taken by trap and net on Georgia mountain summits in 1952. *Ecology* 35: 50–59.

Downes, J. A. 1955. Observations on the swarming flight and mating of *Culicoides* (Diptera: Ceratopogonidae). *Transactions of the Royal Entomological Society of London* 106: 213–236.

———1969. The swarming and mating flight of Diptera. *Annual Review of Entomology* 14: 271–298.

Eisner, T., and F. C. Kafatos. 1962. Defensive mechanisms of arthropods. X. A pheromone promoting aggregation in an aposematic distasteful insect. *Psyche* 69: 53–61.

Gatenby, J. B. 1959. Notes on the New Zealand glow-worm, *Bolitophila* (Arachnocampa) *luminosa. Transactions of the Royal Society of New Zealand* 87: 291–314.

Hölldobler, B. 1976. The behavioral ecology of mating in harvester ants (Hymenoptera: Formicidae: *Pogonomyrmex*). *Behavioral Ecology and Sociobiology* 1: 405–423.

Hölldobler, B., and E. O. Wilson. 1990. *The Ants.* Cambridge, Mass.: Harvard University Press.

———1994. *Journey to the Ants.* Cambridge, Mass.: Harvard University Press.

Hudson, G. V. 1950. *Fragments of New Zealand Entomology.* Wellington, New Zealand: Ferguson and Osborn.

Knab, F. 1906. The swarming of *Culex pipiens. Psyche* 13: 123–133.

Langmuir, I. 1938. The speed of the deer fly. *Science* 87: 233–234.

Linsley, E. G., T. Eisner, and A. B. Klots. 1961. Mimetic assemblages of sibling species of lycid beetles. *Evolution* 15: 15–29.

Meyer-Rochow, V. B., and E. Eguchi. 1984. Thoughts on the possible function and origin of bioluminescence in the New Zealand glowworm *Arachnocampa luminosa* (Diptera: Keroplatidae), based on electrophysiological recordings of spectral responses from the eyes of male adults. *New Zealand Entomologist* 8: 111–119.

Miller, G. 1935. Synchronous firefly flashing. *Science* 81: 590–591.

Neumann, D. 1986. Life cycle strategies of an intertidal midge between subtropic and Arctic latitudes. In F. Taylor and R. Karban (eds.), *The Evolution of Insect Life Cycles.* New York: Springer-Verlag.

Richards, A. M. 1960. Observations on the New Zealand glow worm *Arachnomorpha luminosa* (Skuse) 1890. *Transactions of the Royal Society of New Zealand* 88: 559–574.

Rudinsky, J. A. 1969. Masking the aggregation pheromone in *Dendroctonus pseudotsugae* Hopk. *Science* 166: 884–885.

Rudinsky, J. A., L. C. Ryker, R. R. Michael, L. M. Libbey, and M. E. Morgan. 1976. Sound production in Scolytidae: female sonic stimulus of male pheromone release in two *Dendroctonus* beetles. *Journal of Insect Physiology* 22: 1675–1681.

Tokeshi, M., and K. Reinhardt. 1996. Reproductive behaviour in *Chironomus anthracinus* (Diptera: Chironomidae), with a consideration of the evolution of swarming. *Journal of Zoology, London* 240: 103–112.

Townsend, C. H. T. 1927. On the *Cephenemyia* mechanism and the daylight-day circuit of the earth by flight. *Journal of the New York Entomological Society* 35: 245–252.

Waldbauer, G. P., and J. G. Sternburg. 1973. Polymorphic termination of diapause by cecropia: genetic and geographical aspects. *Biological Bulletin* 145: 627–641.

Walker, B. W. 1952. A guide to the grunion. *California Fish and Game* 38: 409–420.

Myriads of Mayflies

Alba-Tercedor, J., and A. Sanchez-Ortega (eds.). 1991. *Overview and Strategies of Ephemeroptera and Plecoptera.* Gainesville, Fla.: Sandhill Crane Press.

Brittain, J. E. 1982. Biology of mayflies. *Annual Review of Entomology* 27: 119–147.

Carr, J. F., and J. K. Hiltunen. 1965. Changes in the bottom fauna of western Lake Erie from 1930 to 1961. *Limnology and Oceanography* 10: 551–569.

Edmunds, G. F., Jr., S. L. Jensen, and L. Berner. 1976. *The Mayflies of North and Central America*. Minneapolis: University of Minnesota Press.

Ehrlich, P. R., and A. H. Ehrlich. 1972. *Population, Resources, Environment*, 2nd ed. San Francisco: W. H. Freeman.

Fremling, C. R., and D. K. Johnson. 1990. Recurrence of *Hexagenia* mayflies demonstrates improved water quality in Pool 2 and Lake Pepin, Upper Mississippi River. In I. C. Campbell (ed.), *Mayflies and Stoneflies*. Norwell, N.Y.: Kluwer Academic Publishers.

Lyman, F. E. 1944. Notes on emergence, swarming, and mating of *Hexagenia* (Ephemeroptera). *Entomological News* 55: 207–210.

Mason, W. T., Jr., C. R. Fremling, and A. V. Nebeker. 1995. Aquatic insects as indicators of water quality. In E. T. La Roe (ed.), *Our Living Resources*. U.S. Department of the Interior, National Biological Service.

Needham, J. G., J. R. Traver, and Y. C. Hsu. 1935. *The Biology of Mayflies*. Ithaca: Comstock.

Schneider, P. 1997. Clear progress: twenty-five years of the Clean Water Act. *Audubon* 99(5): 36–47.

Sweeney, B. W., and R. L. Vannote. 1982. Population synchrony in mayflies: a predator satiation hypothesis. *Evolution* 36: 810–821.

Ward, J. V. 1992. *Aquatic Insect Ecology*, vol. 1: *Biology and Habitat*. New York: Wiley.

Wood, K. G. 1973. Decline of *Hexagenia* (Ephemeroptera) nymphs in western Lake Erie. In W. L. Peters and J. G. Peters (eds.), *Proceedings of the First International Congress on Ephemeroptera*. Leiden: E. J. Brill.

Swarms of Cicadas

Alexander, R. D., and T. E. Moore. 1962. The evolutionary relationships of 17-year and 13-year cicadas, and three new species (Homoptera, Cicadidae, *Magicicada*). *Miscellaneous Publications, Museum of Zoology, University of Michigan* 121: 1–59.

Cott, H. B. 1957. *Adaptive Coloration in Animals*. London: Methuen.

Dybas, H. S., and D. D. Davis. 1962. A population census of seventeen-year periodical cicadas (Homoptera: Cicadidae: *Magicicada*). *Ecology* 43: 432–444.

Dybas, H. S., and M. Lloyd. 1974. The habitats of 17-year periodical cicadas (Homoptera: *Magicicada* spp.). *Ecological Monographs* 44: 279–324.

Felt, E. P. 1905. *Insects affecting Park and Woodland Trees*. New York State Museum Memoir, no. 8.

Lloyd, M., and H. S. Dybas. 1966. The periodical cicada problem. II. Evolution. *Evolution* 20: 466–505.

Maier, C. T. 1982. Observations on the seventeen-year periodical cicada, *Magicicada septendecim* (Hemiptera: Homoptera: Cicadidae). *Annals of the Entomological Society of America* 75: 14–23.

Marlatt, C. L. 1898. A new nomenclature for the broods of the periodical cicada: miscellaneous results of work of the Division of Entomology. *Bulletin of the United States Department of Agriculture, Division of Entomology* 18: 52–58.

Matthews, R. W., and J. R. Matthews. 1978. *Insect Behavior.* New York: John Wiley and Sons.

Myers, J. G. 1929. *Insect Singers.* London: George Routledge and Sons.

Oldenburg, H. 1666. Some observations of swarms of strange insects and the mischiefs done by them. *Philosophical Transactions of the Royal Society of London.* 1: 137–138.

Simmons, J. A., E. G. Wever, and J. M. Pylka. 1971. Periodical cicada: sound production and hearing. *Science* 171: 212–213.

Stannard, L. J., Jr. 1975. *The Distribution of Periodical Cicadas in Illinois.* Illinois Natural History Survey Biological Note, 91 (Urbana).

When Chicago braced for the onslaught of the 17-year cicada. 1990. *The Food Insects Newsletter* 3: 3, 5.

White, J. 1980. Resource partitioning by ovipositing cicadas. *American Naturalist* 115: 1–28.

White, J., and M. Lloyd. 1975. Growth rates of 17- and 13-year periodical cicadas. *American Naturalist* 94: 127–143.

White, J., and C. E. Strehl. 1978. Xylem feeding by periodical cicada nymphs on tree roots. *Ecological Entomology* 3: 323–327.

Wigglesworth, V. B. 1972. *The Principles of Insect Physiology.* London: Chapman and Hall.

Williams, K. S., and C. Simon. 1995. The ecology, behavior, and evolution of periodical cicadas. *Annual Review of Entomology* 40: 269–295.

Williams, K. S., K. G. Smith, and F. M. Stephen. 1993. Emergence of 13-year periodical cicadas (Cicadidae: *Magicicada*): phenology, mortality, and predator satiation, *Ecology* 74: 1143–1152.

Associating with Other Species

Bayliss, M., and N. E. Pierce. 1993. The effects of ant mutualism on the foraging and diet of lycaenid caterpillars. In N. E. Stamp and T. M. Casey (eds.), *Caterpillars.* New York: Chapman and Hall.

DeVries, P. J. 1990. Enhancement of symbiosis between butterfly caterpillars and ants by vibrational communication. *Science* 248: 1104–1106.

Fiedler, K., and U. Maschwitz. 1987. Functional analysis of the myrmecophilous relationships between ants (Hymenoptera: Formicidae) and lycaenids (Lepidoptera: Lycaenidae). *Zoologische Beiträge* 31: 409–416.

Hölldobler, B., and E. O. Wilson. 1994. *Journey to the Ants.* Cambridge, Mass.: Harvard University Press.

Linsley, E. G., T. Eisner, and A. B. Klots. 1961. Mimetic assemblages of sibling species of lycid beetles. *Evolution* 15: 15–29.

Metcalf, R. L., and R. A Metcalf. 1993. *Destructive and Useful Insects.* New York: McGraw-Hill.

Pierce, N. E., and S. Easteal. 1986. The selective advantage of attendant ants for the larvae of a lycaenid butterfly, *Glaucopsyche lygdamus. Journal of Animal Ecology* 55: 451–462.

Pierce, N. E., R. L. Kitching, R. C. Buckley, M. F. J. Taylor, and K. F. Benbow. 1987. The costs and benefits of cooperation between the Australian lycaenid butterfly, *Jalmenus evagoras,* and its attendant ants. *Behavioral Ecology and Sociobiology* 21: 237–248.

Sullivan, K. A. 1984. Information exploitation by downy woodpeckers in mixed-species flocks. *Behaviour* 91: 294–311.

Tutt, J. W. 1900. *British Lepidoptera,* vol. 2. London: Swan Sonnenschein and Co.

Wheeler, W. M. 1910. *Ants: Their Structure, Development and Behavior.* New York: Columbia University Press.

Acknowledgments

I am greatly indebted to the many friends and colleagues who helped me by giving generously of their time, advice, and knowledge: Marianne Alleyne, May Berenbaum, Julia Berger, Sam Beshers, Guy Bloch, John Bouseman, Lincoln Brower, William Callahan, Patrice Charlebois, Lynda Corkum, Susan Fahrbach, Terrence Fitzgerald, Calvin Fremling, Arthur Ghent, Fred Gottheil, Bert Hall, Larry Hanks, Brian Harrington, Edwin Jahiel, Rabbi Norman Klein, James Krysan, Richard Leskosky, James Lloyd, Thomas Moore, James Nardi, Robert Novak, Sandra Perez, Scott Robinson, David Seigler, Carl Schuster, Jr., James Sternburg, Orley Taylor, Joan Walsh, David Sloan Wilson, Edward O. Wilson, and David Wood.

My wife, the late Stephanie Waldbauer, gave me constant encouragement and read and criticized the entire manuscript. Katherine Brown-Wing did the illustrations that so nicely complement the text. Special thanks are due to Dorothy Nadarski, who patiently typed and retyped the manuscript. The book benefited greatly from the constant encouragement of Michael Fisher and Ann Downer-Hazell, the careful editing of Nancy Clemente, and the matchless design talents of Marianne Perlak.

Index